Molluscan Communities of the Florida Keys and Adjacent Areas

Molluscan Communities of the Florida Keys and Adjacent Areas

THEIR ECOLOGY AND BIODIVERSITY

Edward J. Petuch

Florida Atlantic University, Boca Raton, USA

Robert F. Myers

Coral Graphics/Seaclicks, Wellington, Florida, USA

CRC Press
Taylor & Francis Group
Boca Raton London New York

CRC Press is an imprint of the
Taylor & Francis Group, an **informa** business

CRC Press
Taylor & Francis Group
6000 Broken Sound Parkway NW, Suite 300
Boca Raton, FL 33487-2742

First issued in paperback 2020

ISBN-13: 978-1-4822-4918-7 (hbk)
ISBN-13: 978-0-367-65891-5 (pbk)

Library of Congress Cataloging-in-Publication Data

Petuch, Edward J.
  Molluscan communities of the Florida Keys and adjacent areas : their ecology and biodiversity / Edward J. Petuch and Robert F. Myers.
    pages cm
  Includes bibliographical references and index.
  ISBN 978-1-4822-4918-7
  1. Mollusks--Florida--Florida Keys. 2. Mollusks--Ecology--Florida--Florida Keys. I. Myers, Robert F., 1953- II. Title.

  QL415.F6P48 2014
  594.09759'41--dc23   2014021177

Visit the Taylor & Francis Web site at
http://www.taylorandfrancis.com

and the CRC Press Web site at
http://www.crcpress.com

Dedication

To Linda Petuch, Eric Petuch, Brian Petuch, and Jennifer Petuch and to Patrice Marker, Laura Myers, and Robbie Myers

and

to the memory of Rachel Carson

Contents

Large water spout forming off Missouri Key.

Acknowledgments

For assisting in the gathering of data in the field and for collecting study specimens in the Florida Keys and Lake Worth, we thank the following individuals: Nelson Jimenez, Middle Torch Key, Florida; Eddie Matchett, Okeechobee, Florida; Clifford Swearingen, Fort Lauderdale, Florida; Emyr Foxhall, Cardiff, Wales (Caerdydd, Cymru); Dr. Ron and Mrs. Mary Jo Bopp, Sarasota, Florida; Stephen Tressel, Jupiter, Florida; Nicolas Mazzoli, Pine Island, Florida; Shawn Webster, Conch Key, Florida; William Bennight, Jacksonville, North Carolina; Mark Geiser, Winchester, Kentucky; Everett Long, Cedar Point, North Carolina; Douglas Jeffrey, Key Largo, Florida; Robert Pace, Miami, Florida; James Lauriello, Stuart, Florida; Dr. M.G. Harasewych, Smithsonian Institution, Washington, D.C.; Carole Marshall, West Palm Beach, Florida; Ida Bucheck, Singer Island, Florida; Jennie Petuch, Tampa, Florida; Michael Black, West Palm Beach, Florida; Kevan Sunderland, Plantation Key, Florida; Brian Morgan, Cudjoe Key, Florida; Günther Herndl, Vienna, Austria; the late Theodore Kalafut, Largo, Florida; and Dr. Howard Peters, University of York, UK.

For guiding the senior author to the best-developed vermetid reefs of the Ten Thousand Islands, we thank A. Kenneth ("Kenny") Brown, Jr., Chokoloskee, Florida. For the donation of valuable research specimens, many of which are illustrated in this book, we thank the following: Thomas and Paula Honker, Delray Beach, Florida; Andrew Dickson, Miami, Florida; Caitlin Hanley, Florida Atlantic University; Pierre Recourt, Egmond aan Zee, The Netherlands; Jeff Whyman, Boynton Beach, Florida; Dr. Kenneth and Mrs. Wendy Keaton, Lauderhill, Florida; Jodilynn Duggan, Florida Atlantic University; and also Everett Long, Carole Marshall, Dr. M. G. Harasewych, and James Lauriello. Special thanks go to Robert Owens, Boca Raton, Florida, for his invaluable assistance to the senior author while conducting surveys of the molluscan fauna of the Ten Thousand Islands.

For many of the locality shots and live specimen photographs used in this book, we thank Eddie Matchett, Dr. Howard Peters, Dr. Ron Bopp, Robert Owens, and William Bennight. For many of the beautiful photographs of Florida Keys shells illustrated here, we give special thanks to Dennis M. Sargent, Mount Dora, Florida. Thanks also to Dr. Manuel Tenorio, University of Cadiz, Spain, for extracting and photographing the radulae of *Gradiconus burryae* and *Gradiconus mazzolii*. Special thanks also go to Dr. M. G. Harasewych (Division of Mollusks, Smithsonian Institution), and Dr. Thomas A. Frankovich (Florida Bay Interagency Science Center, Key Largo) for reviewing the manuscript and offering many helpful suggestions; to Carole Marshall, West Palm Beach Florida, for proofreading the manuscript; and to Dr. Tobin Hindle, Florida Atlantic University, for technical assistance in setting up the illustration files. Thanks also go to John Sulzycki, senior editor for life sciences, and Marsha Hecht, project editor, CRC Press, for their valuable insights into the technical aspects of publishing this book.

View of the shoreline of southernmost Lower Matecumbe Key, Middle Florida Keys, showing Indian Key in the distance. (Photograph by Eddie Matchett)

Introduction

Endemism and biodiversity in the Florida Keys and adjacent areas

Long admired for their exotic tropical beauty, the island chains of the Florida Keys constitute one of the great natural wonders of North America. This extended archipelago, curving outward into the Gulf of Mexico, contains the only set of tropical marine environments found anywhere in the continental United States. Anyone visiting this area cannot help but notice that the shimmering turquoise and emerald waters of the Florida Keys literally vibrate with life. This sense of motion was aptly described by the great marine naturalist Rachel Carson, who stated, "In the multicolored sea gardens seen from a boat as one drifts above them, there is a tropical lushness and mystery, a throbbing sense of the pressure of life; in coral reef and mangrove swamp" (1955, p. 191). Each environment of the Florida Keys is filled with swarming multitudes of marine organisms, all woven together into some of the most intricate ecosystems seen anywhere in the Atlantic Ocean.

Unlike other localities along the coastlines of the eastern United States and Caribbean, the Florida Keys archipelago is the only area known to harbor ecosystems composed of organisms from two different oceanographic and climatological regimes. Within the shallow marine communities of the Florida Keys, many of the resident organisms are known to be migrants from the cooler, temperate waters of the Carolinas and Georgia to the north, while others are migrants from the tropical waters of the Caribbean Sea to the south. This interweaving of two distinctly different marine biotas is further enhanced by a large suite of endemic species: those that have evolved in, and are restricted to, the Florida Keys area. The overlap of these three biotic elements, the warm-tolerant northern species, the cool-tolerant southern species, and the restricted Keys endemics, has created characteristic southern Florida marine communities that stand out from all the others of the world's tropical zones. This unusually rich composite marine world extends outward from the Florida Keys island chain and is found from the Ten Thousand Islands of Collier and Monroe Counties, to the islands of the Dry Tortugas, to Florida Bay and Biscayne Bay, and northward to the deep reefs and coastal lagoons of Palm Beach County.

Of all the types of marine organisms found throughout this extended area, the phylum Mollusca predominates, with individual gastropod and bivalve species often carpeting entire seafloors or shorelines. This amazing abundance of individuals occurs in all the habitat types seen throughout the Florida Keys and adjacent areas, including coral reefs, sea grass beds, open sand bottoms, and mangrove forests. As demonstrated by the biodiversity analyses in Chapters 2 through 9 of this book, the combined number of species from all of these habitats, both macromollusks and micromollusks, approaches 1500 individual taxa. This large number of species is equivalent to the species richness seen on many of the South Pacific islands, underscoring the special nature of the southern Florida malacofauna. As the most species-rich group of animals, the mollusks of the Florida Keys area take on a special prominence, acting as a "watch dog" for the overall health of the resident marine ecosystems.

The unusual composite nature of the Florida Keys molluscan fauna is the direct result of the regional geography, with the island chains projecting into tropical waters and at the same time contiguous with cooler continental areas to the north. The unique habitats that were created by this blending of oceanographic conditions, along with their resident malacofaunas, have led to the evolution of a distinct marine biogeographical entity known as the *Floridian Molluscan Subprovince* (extending from Cape Sable to Florida Bay and from the Dry Tortugas to Biscayne Bay; see Petuch, 2013: 47–57, for the quantitative analysis and description of the Floridian malacofauna). As a subprovince, this southern Floridian biogeographical unit is considered to be a subdivision of the much larger *Carolinian Molluscan Province*, which encompasses the coastline of the entire southeastern United States and Gulf of Mexico, from Cape Hatteras, North Carolina, around the Florida Keys, to Texas and Yucatan, Mexico. Of the five subprovinces of the Carolinian Molluscan Province (see Petuch, 2013: 15–20, 31–74, for the quantitative analyses of the province and subprovinces), the Floridian Molluscan Subprovince alone contains faunal elements derived from two adjacent subprovinces: the *Georgian Molluscan Subprovince* (extending from Cape Hatteras to Palm Beach) and the *Suwannean Molluscan Subprovince* (extending from Cape Sable, Florida, northward to the Mississippi River Delta). A general Carolinian Molluscan Province influence is also present within the Floridian Subprovince in the form of a large complement of widespread species that range from Yucatan, Mexico, all the way to Cape Hatteras, North Carolina. These temperate water northern-derived taxa occur together with widespread high tropical mollusks from the *Caribbean Molluscan Province*, most of which range from the Bahamas and Bermuda south to northern South America (see Petuch, 2013). The wide-ranging Carolinian and Caribbean components, along with the Florida Keys endemics, produce a biogeographically hybridized malacofauna. The lists that follow give examples of some of the more prominent species that occur within the composite molluscan fauna of the Florida Keys and adjacent areas. These gastropods and bivalves come from a wide bathymetric range, and these lists reflect only their biogeographical distributions (for the definitions and descriptions of western Atlantic molluscan faunal provinces and subprovinces as used here, see Petuch, 2013). Some of the more well known of these include the following:

FAUNAL ELEMENTS FROM THE CAROLINIAN MOLLUSCAN PROVINCE
(wide-ranging species found from Cape Hatteras to the Florida Keys and the entire Gulf of Mexico as far as Isla Contoy, Yucatan Peninsula, Mexico; not found in the Caribbean region)
Gastropoda
Cypraeidae
Macrocypraea (Lorenzicypraea) cervus (Linnaeus, 1771) (Figure 7.5G, H; Chapter 7)
Cassidae
Cassis spinella Clench, 1944 (Figure 6.6G, H; Chapter 6)
Strombidae
Strombus alatus Gmelin, 1791 (Figure 6.6E, F; Chapter 6)
Muricidae
Chicoreus dilectus (A. Adams, 1855) (Figure 4.14D–F; Chapter 4)
Fasciolariidae

Cinctura hunteria (Perry, 1811) (Figure 4.4C, D; Chapter 4)

Triplofusus papillosus (Sowerby I, 1825) (Figure 4.10H; Chapter 4)

Busyconidae

Sinistrofulgur sinistrum (Hollister, 1958) (ranges from North Carolina to Louisiana) (Figure 6.3E; Chapter 6)

Conidae

Dauciconus amphiurgus (Dall, 1889) (Figure 8.5G, H; Chapter 8)

Conasprelloides stimpsoni (Dall, 1902) (Figure 9.7A; Chapter 9)

Lindaconus atlanticus (Clench, 1942) (Figure Figure 6.9E, F; Chapter 6)

Conilithidae

Kohniconus delessertii (Recluz, 1843) (Figure 9.2F; Chapter 9)

Bivalvia

Pectinidae

Argopecten irradians taylorae Petuch, 1987 (ranges from southern Biscayne Bay to Alabama) (Figure 4.12A–D; Chapter 4)

Nodipecten fragosus (Conrad, 1849) (Figure 6.7L; Chapter 6)

Veneridae

Mercenaria campechiensis (Gmelin, 1791) (ranges from North Carolina to Yucatan, Mexico) (Figure 6.2E; Chapter 6)

FAUNAL ELEMENTS FROM THE CARIBBEAN MOLLUSCAN PROVINCE (wide-ranging species found from the Bahamas and Bermuda southward to the Amazon River Mouth; also southeastern Florida as far north as Palm Beach)

Gastropoda

Neritidae

Nerita (Linnerita) peloronta Linnaeus, 1758 (Figure 3.6E, F; Chapter 3)

Nerita versicolor Gmelin, 1791 (Figure 3.6A, B; Chapter 3)

Littorinidae

Cenchritis muricatus (Linnaeus, 1758) (Figure 3.3A, B; Chapter 3)

Strombidae

Eustrombus gigas (Linnaeus, 1758) (Figure 6.10H, I; Chapter 6)

Lobatus raninus (Gmelin, 1791) (Figure 4.8K; Chapter 4)

Strombus pugilis Linnaeus, 1758 (Figure 6.6C, D; Chapter 6)

Cassidae

Cassis flammea (Linnaeus, 1758) (Figure 7.7A, B; Chapter 7)

Cassis madagascariensis Lamarck, 1822 (Figure 6.6I, J; Chapter 6)

Casmaria atlantica Clench, 1944 (Figure 6.4E; Chapter 6)

Ovulidae

Cyphoma gibbosum (Linnaeus, 1758) (Figure 5.6A–C; Chapter 5)

Cyphoma signatum Pilsbry and McGinty, 1939 (Figure 5.6I; Chapter 5)

Ranellidae

Cymatium (Ranularia) rehderi Verrill, 1950 (Figure 8.3E, F; Chapter 8)

Muricidae

Chicoreus mergus E. Vokes, 1974 (Figure 8.2I; Chapter 8)

Phyllonotus oculatus (Reeve, 1845) (Figure Figure 8.2G, H; Chapter 8)

Conidae

Attenuiconus attenuatus (Reeve, 1844) (Figure 6.9B; Chapter 6)

Atlanticonus granulatus (Linnaeus, 1758) (Figure 8.3J; Chapter 8)

Dauciconus daucus (Hwass, 1792) (Figure 6.9A; Chapter 6)

Gladioconus mus (Hwass, 1792) (Figure 7.10K; Chapter 7)
Stephanoconus regius (Gmelin, 1791) (Figure 7.10J; Chapter 7)
Bivalvia
Pectinidae
Argopecten nucleus (Born, 1778) (Figure 4.12E, F; Chapter 4)
Lindapecten exasperatus (Sowerby II, 1842) (Figure 4.12G, H; Chapter 4)

SPECIES ENDEMIC TO THE FLORIDIAN MOLLUSCAN SUBPROVINCE (taxa restricted to the Florida Keys, Florida Bay, and Biscayne Bay; some also along Palm Beach)
Gastropoda
Calliostomatidae
Calliostoma adelae Schwengel, 1951 (Figure 4.8D; Chapter 4)
Turbinidae
Lithopoma americana (Gmelin, 1791) (Figure 4.8A; Chapter 4)
Modulidae
Modulus calusa Petuch, 1988 (Figure 4.7A–F; Chapter 4)
Ovulidae
Cyphoma rhomba Cate, 1978 (Figure 5.6G, H; Chapter 5)
Cyphoma sedlaki Cate, 1976 (Figure 5.6J; Chapter 5)
Muricidae
Favartia pacei Petuch, 1988 (Figure 5.2G; Chapter 5)
Murexiella caitlinae Petuch and Myers, new species (Figure 5.2L, J; Chapter 5)
Murexiella kalafuti Petuch, 1987 (Figure 9.1G; Chapter 9)
Fasciolariidae
Cinctura tortugana (Hollister, 1957) (Figure 9.1L; Chapter 9)
Leucozonia jacarusoi Petuch, 1978 (Figure 9.1K; Chapter 9)
Melongenidae
Melongena (Rexmela) bicolor (Say, 1827) (Figure 3.10I, J; Chapter 3)
Buccinidae
Hesperisternia sulzyckii Petuch and Myers, new species (Figure 9.3E, F; Chapter 9)
Nassariidae
Uzita swearingeni Petuch and Myers, new species (Figure 5.3J–L; Chapter 5)
Uzita websteri Petuch and Sargent, 2011 (Figure 5.3G, H; Chapter 5)
Columbellidae
Zafrona taylorae Petuch, 1987 (Figure 4.12B; Chapter 4)
Cancellariidae
Cancellaria adelae Pilsbry, 1940 (Figure 6.8I; Chapter 6)
Olividae
Americoliva matchetti Petuch and Myers, new species (Figure 9.6D–F; Chapter 9)
Americoliva recourti Petuch and Myers, new species (Figure 9.6A–C; Chapter 9)
Volutidae
Clenchina dohrni (Sowerby III, 1903) (Figure 9.5G, H; Chapter 9)
Rehderia schmitti (Bartsch, 1931) (Figure 9.5C, D; Chapter 9)
Scaphella junonia elizabethae Petuch and Sargent, 2011 (Figure 6.8G, H; Chapter 6)
Marginellidae
Prunum frumari Petuch and Sargent, 2011 (Figure 9.4K; Chapter 9)
Conidae
Gradiconus anabathrum tranthami (Petuch, 1995) (Figure 6.10A–D; Chapter 6)
Gradiconus burryae (Clench, 1942) (Figure 4.9A–F; Chapter 4)

Gradiconus mazzolii Petuch and Sargent, 2011 (Figure 5.4F–K; Chapter 5)
Conilithidae
Jaspidiconus fluviamaris Petuch and Sargent, 2011 (Figure 6.9G, H; Chapter 6) (also
    Palm Beach)
Jaspidiconus pealii (Green, 1830) (Figure 4.9G–L; Chapter 4)
Jaspidiconus vanhyningi (Rehder, 1944) (Figure 6.9I, J; Chapter 6) (also Palm Beach)
Drilliidae
Cerodrillia clappi Bartsch and Rehder, 1939 (Figure 5.2D; Chapter 5)
Bullidae
Bulla frankovichi Petuch and Sargent, 2011 (Figure 5.4D, E, L; Chapter 5)
Haminoeidae
Haminoea taylorae Petuch, 1987 (Figure 4.2F, G; Chapter 4) (also Palm Beach)
Bivalvia
Arcidae
Arca rachelcarsonae Petuch and Myers, new species (Figure 8.4A–C; Chapter 8)
Pectinidae
Caribachlamys mildredae (Bayer, 1941) (Figure 7.9H, I; Chapter 7) (also Palm Beach)
Chamidae
Chama inezae (Bayer, 1943) (Figure 7.11C; Chapter 7) (also Palm Beach)

For the ecological and biodiversity analyses given in each of the following chapters, only the species in macromollusk families were utilized (gastropod and bivalve families with species that have average lengths greater than 5 mm). For the Gastropoda, these marine macromollusk groups include 86 families: Pleurotomariidae, Haliotidae, Lottiidae, Fissurellidae, Solariellidae, Margaritidae, Calliostomatidae, Trochidae, Turbinidae, Liotiidae, Phasianellidae, Neritidae, Phenacolepidae, Littorinidae, Planaxidae, Litiopidae, Truncatellidae, Potamididae, Cerithiidae, Batillariidae, Turritellidae, Siliquariidae, Vermetidae, Modulidae, Strombidae, Xenophoridae, Epitoniidae, Nystiellidae, Janthinidae, Architectonicidae, Calyptreidae, Capulidae, Hipponicidae, Vanikoridae, Mathildidae, Capulidae, Cypraeidae, Lamellariidae, Triviidae, Pediculariidae, Naticidae, Tonnidae, Cassidae, Ficidae, Ranellidae, Bursidae, Personidae, Muricidae, Fasciolariidae, Melongenidae, Busyconidae, Buccinidae, Nassariidae, Colubrariidae, Columbellidae, Harpidae, Turbinellidae, Costellariidae, Mitridae, Volutomitridae, Volutidae, Olivellidae, Olividae, Marginellidae, Cystiscidae, Cancellariidae, Conidae, Conilithidae, Terebridae, Turridae, Cochlespiridae, Strictispiridae, Clathurellidae, Crassispiridae, Horaiclavidae, Drilliidae, Raphitomidae, Zonulispiridae, Amathinidae, Acteonidae, Bullidae, Haminoeidae, Cylichnidae, Aplustridae, Ellobiidae, and Siphonariidae.

For the Bivalvia, these marine macromollusk groups include 54 families: Nuculidae, Solemyidae, Nuculanidae, Yoldiidae, Arcidae, Noetiidae, Glycymeridae, Mytilidae, Pteriidae, Isognomatidae, Malleidae, Ostreidae, Gryphaeidae, Pinnidae, Limidae, Limopsidae, Pectinidae, Propeamussiidae, Spondylidae, Plicatulidae, Anomiidae, Crassitellidae, Astartidae, Carditidae, Pandoridae, Lyonsiidae, Periplomatidae, Thraciidae, Verticordiidae, Poromyidae, Cuspidariidae, Lucinidae, Ungulinidae, Thyasiridae, Chamidae, Lasaeidae, Hiatellidae, Gastrochaenidae, Trapezidae, Sportellidae, Corbiculidae, Cardiidae, Veneridae, Tellinidae, Donacidae, Psammobiidae, Semelidae, Solecurtidae, Pharidae, Mactridae, Dreissenidae, Myidae, Corbulidae, and Pholadidae.

The micromollusks (gastropod and bivalve families with species that average less than 5 mm in length) of the Florida Keys and adjacent areas are still poorly studied, and their systematics and taxonomy are still largely in a state of flux. Because of this, the many families that fall into this size category are not covered in this book and are not used for any ecological or biodiversity analyses. These tiny shells are abundant in many environments, and they constitute a large part of the total malacofauna, with as many as 300 species possibly present between all the gastropod and bivalve families. Future studies will doubtlessly bring to light many new species of micromollusks in the Keys area. For the Gastropoda, these marine micromollusk groups include 30 families: Scissurellidae, Chilodontidae, Rissoidae, Rissoinidae, Rissoellidae, Seguenziidae, Skeneidae, Skeneopsidae, Omalogyridae, Vitrinellidae, Torniidae, Caecidae, Cerithiopsidae, Alabidae, Triphoridae, Melanellidae, Stiliferidae, Aclididae, Atlantidae, Carinariidae, Vanikoroidae, Fossaridae, Pyramidellidae, Odostomiidae, Eulimidae, Tubonillidae, Ringiculidae, Philinidae, Retusidae, and Volvatellidae.

For the Bivalvia, these marine micromollusk groups include only four families: Manzanellidae, Philobryidae, Condylocardiitidae, and Spheniopsidae.

Other molluscan groups, such as the classes Cephalopoda, Scaphopoda, Polyplacophora, and Aplacophora (including Caudofoveata and Solenogastres), are not covered in this book; the soft-bodied nudibranch gastropod "sea slugs," the teredinid "shipworm" bivalves, and shell-less pelagic groups also are not covered. As in the case of the micromollusks, the systematics, taxonomy, and life histories of many of these animals are still poorly known, and these groups are not included in any of the subsequent ecological and biodiversity analyses. Even without the micromollusks, chitons, tusk shells, and soft-bodied groups, the macromollusks still provide the greatest insight into the workings of molluscan ecosystems and the richness of molluscan faunas. As determined by Mikkelsen and Bieler (2008), the 54 macrobivalve families found in the Florida Keys area contain, altogether, 364 species (of a total of 375 bivalve species in 58 families). This relatively small number of bivalves contrasts greatly with the over 850 species found in the 85 families of macrogastropods. Extrapolating on these patterns of species richness, further research in the Florida Keys area, especially the deep reef talus slopes and deep terraces, will surely uncover many more new and rarely seen species of mollusks. A list of the macromollusks of the Florida Keys and adjacent areas, arranged systematically, is given at the end of this book.

This book is meant to be simply a starting point for molluscan ecological studies of the coral reef tracts, mangrove forests, and lagoon areas of the American biological treasure known as the Florida Keys. It is hoped the ecological classification framework that is presented in the following chapters will serve as a foundation for future marine ecosystem studies in the Florida Keys and Ten Thousand Islands and along the Palm Beach coast. By including color illustrations of over 500 different species, subspecies, and forms, this book also serves as a conchological field guide for shell collectors, divers, and naturalists. The mystery and beauty of the Florida Keys and adjacent areas, with their spectacular molluscan faunas, was eloquently captured by Rachel Carson in *The Edge of the Sea* (1955, p. 247), where she stated: "And as the years pass, and as the centuries merge into the unbroken stream of time, these architects of coral reef and mangrove swamp build toward a shadowy future. But neither the corals nor the mangroves, but the sea itself will determine when that which they build will belong to the land, or when it will be reclaimed for the sea." For now, these unique American environments belong to the marine world, and they offer years of excitement and discovery right here on our own doorstep.

About the Authors

Edward J. Petuch was born in Bethesda, Maryland, in 1949. Raised in a Navy family, he spent many of his childhood years collecting living and fossil shells in such varied localities as Chesapeake Bay, California, Puerto Rico, Panama, Cuba, and Wisconsin. His early interests in malacology and oceanography eventually led to BA and MS degrees in zoology from the University of Wisconsin–Milwaukee. While in Wisconsin, his thesis work concentrated on the molluscan biogeography of West Africa. There, he collected mollusks and traveled extensively in Morocco, Western Sahara, the Canary Islands, Senegal, Gambia, Sierra Leone, Ivory Coast, and the Cameroons. At this time, he also made frequent research trips to both coasts of Mexico and the Great Barrier Reef of Belize.

Continuing his education, Petuch studied marine biogeography and malacology under Gilbert Voss and Donald Moore at the Rosenstiel School of Marine and Atmospheric Sciences at the University of Miami. During this time, his dissertation work involved intensive collecting and fieldwork (often on shrimp boats) in Colombia, Venezuela, Barbados, the Grenadines, Curacao, and Brazil. After receiving his PhD in oceanography in 1980, Petuch undertook 2 years of National Science Foundation-sponsored postdoctoral research on molluscan paleoecology and biogeography with Geerat Vermeij at the University of Maryland. While there, he also held a research associateship with the Department of Paleobiology at the National Museum of Natural History at the Smithsonian Institution (under the sponsorship of Thomas Waller) and conducted fieldwork on the Plio-Pleistocene fossil beds of Florida and North Carolina and the Miocene of Maryland and Virginia.

Petuch has also collected and studied living mollusks in Australia, Papua New Guinea, the Fiji Islands, French Polynesia, Japan, the Mediterranean coasts of North Africa and Spain, the Bahamas, Nicaragua, Costa Rica, and Uruguay. This research has led to the publication of over 150 scientific papers and the descriptions of over 1000 new species of mollusks and over 60 new genera. His previous 15 books are well-known research texts within the malacological and paleontological communities, and some of the better known include *Cenozoic Seas: The View from Eastern North America* (2004); *The Geology of the Everglades and Adjacent Areas* (with Charles Roberts, 2007); *Molluscan Paleontology of the Chesapeake Miocene* (with Mardie Drolshagen, 2010); *Biodiversity and Biogeography of Western Atlantic Mollusks* (2013); *New Caribbean Molluscan Faunas* (1987); *Atlas of Florida Fossil Shells* (1994); *Rare and Unusual Shells of Southern Florida* (with Dennis Sargent, 2011); and *Atlas of the Living Olive Shells of the World* (with Dennis Sargent, 1986).

Currently, Petuch is a professor of geology in the Department of Geosciences at Florida Atlantic University in Boca Raton, where he teaches courses on oceanography, paleontology, and physical geology. When not collecting and studying mollusks in the Florida Keys or the nearby Lake Worth Lagoon, Petuch leads an active career as a musician, giving regular concerts on the pipe organ and playing the recorder and harpsichord in university chamber ensembles.

Robert F. Myers was born in 1953 and spent much of his childhood in southeast Asia, where he developed an intense interest in natural history. His diving and passion for marine life took hold during his early high school years in Hong Kong and continued at the University of Hawaii. There, he assisted well-known ichthyologist John E. Randall by collecting fishes for both ciguatera research and fish population surveys. After earning a BA degree in zoology, Myers moved to Guam to pursue graduate studies and underwater photography. There, he earned an MS degree in biology from the University of Guam Marine Laboratory, worked as a fisheries biologist for the government of Guam, and founded Coral Graphics.

Myers retired from the government in 1995 and now pursues photography, writing, research, and consulting. He has written or coauthored numerous scientific papers and popular articles on western Pacific fishes and several books. Some of these include *Micronesian Fishes* (three editions, 1989–1999); *Coral Reef Fishes* (with E. Lieske, 1993); *Coral Reef Guide Red Sea* (with E. Lieske, 2004); and *Dangerous Marine Animals* (with M. Bergbauer and M. Kirschner, 2009). He has also developed an application for iPhones and iPads on Florida-Caribbean reef fish identification. Now residing in Wellington, Florida, Myers is currently working on a field guide to Florida and Caribbean marine life and serves on the coral reef fishes Species Survival Commission (SSC) of the International Union for the Conservation of Nature (IUCN).

chapter one

Marine ecosystems of the Florida Keys, Florida Bay, and the Ten Thousand Islands: past and present

Introduction

The tropical marine biosphere of the extreme southern coast of the Florida Peninsula occupies the most diverse set of oceanic habitats found anywhere in the United States (Figure 1.1). These environments are an amalgamation of four ecological regimes (Estuarine, Nearshore Marine, Neritic, and Oceanic) and 10 separate ecological macrohabitats (discussed further in this chapter) and encompass the region of the South Florida Bight (a new geographical term described later in this chapter), Florida Keys and its coral reef tracts, Florida Bay, the Dry Tortugas, Biscayne Bay, and the Ten Thousand Islands (Figure 1.2). Elements of these tropical marine environments also extend northward along the southeastern Florida coast to Palm Beach, where a rich and highly endemic invertebrate fauna occurs on deep reefs and in coastal lagoons and inlets. All of these areas together form a framework of habitats that has allowed for the evolution of uniquely North American marine ecosystems, unlike those found anywhere else on Earth.

Molluscan ecological research in the Florida Keys and adjacent areas

Although the ecosystems and biodiversity of the Florida Keys, Dry Tortugas, and Florida Bay have been studied for nearly 200 years, the molluscan ecology of these areas is still virtually unexplored. Since the 1960s, marine ecologists have concentrated primarily on the ecology of sea grass beds (Dawes, 1987; Frankovich and Zieman, 1994); coral reefs (Dunstan, 1985; Goldberg, 1973; Jaap, 1984; Shinn et al., 1989); and mangrove forests (Davis, 1940), and only preliminary molluscan ecological surveys have ever been conducted. Some of these included brief descriptions of local bivalve ecology (Dame, 2011; Mikkelsen and Bieler, 2008); the geology of the vermetid gastropod reefs of the Ten Thousand Islands (Shier, 1969); and the general molluscan ecology of the mangrove forests (Coomans, 1969). Updated systematic and taxonomic lists, including aspects of molluscan ecology, have also been published within the past two decades. Primary among these were the species lists given by Lyons and Quinn (1995) for the mollusks of the Florida Keys Marine Sanctuary and Levy et al. (1996) for the infaunal and epifaunal mollusks of Florida Bay and the Florida Keys. The taxonomy used in both of these major works, however, is now outdated, and the species lists are incomplete, lacking many of the endemic taxa that are now known to occur in these areas.

Broad overviews of the general community types of the Florida Keys and Florida Bay, such as exposed tidal flats, mangrove forests, sea grass beds, and coral reefs, also have been given by several authors over the past two decades. Principal among these are the studies of the functional ecology and local ecosystem trophodynamics undertaken by Chiappone and Sullivan (1994, 1996), Porter (2001), Roessler et al. (2002) (for southern Biscayne Bay and Card Sound), and Rudnick et al. (2005). These works together offered the first conceptual ecological models for Florida Bay and the Florida Keys reef tract. Details of the biogeochemical interactions of lucinid bivalves in Turtle Grass beds were also recently given by Reynolds et al. (2007), as was a small, concentrated faunal survey of the mollusks found within a single sediment core taken in Florida Bay (Trappe and Brewster-Wingard, 2013). Much of this new data expanded on and enhanced the earlier community and biotope studies undertaken by Voss and Voss (1955), Zischke (1973), and Turney and Perkins (1972). Interestingly, this last work was the only one to focus on the entire molluscan fauna, giving descriptions of habitats, faunal distributions, and ecological limiting factors. Much of the molluscan taxonomy used in Turney and Perkins's 1972 work, however, is now outdated, and the authors did not attempt to quantify the overall molluscan biodiversity of the Florida Keys and Florida Bay. Using mollusks as paleoenvironmental indicators, the shifting paleosalinities and paleoclimates of latest Pleistocene and Holocene Florida Bay were recently studied in detail by Brewster-Wingard and Ishman (1999), Brewster-Wingard et al. (2001), and Wingard and Hudley (2011). Because the samples (mostly sediment cores) used in these studies were restricted to only softbottom areas near sea grass beds, the molluscan faunas that they encountered were limited in species richness, containing only around 42 species of bivalves and gastropods.

The senior author's work in the Florida Keys and extreme southern Florida has focused primarily on the systematics, biodiversity, and biogeography of the gastropods and bivalves. To date, 60 new species of mollusks, collected on the reef tract, sea grass beds, sponge bioherms, coastal lagoons of Palm Beach and off the Dry Tortugas, have been described in a series of books and monographic treatments (including Petuch 1987, 1988, 2004, 2013, and Petuch and Sargent 2011b, 2011c). An important new endemic species radiation of cone shells was also published recently by Petuch and Sargent (2011a), which included the description of a new species of *Gradiconus* from the sponge bioherms off the Lower Florida Keys (see Chapter 5). Broad overviews of the main biotopes that support molluscan communities, including those of the coral reefs, sea grass beds, and deepwater coralline algal beds, were given by Petuch and Sargent (2011b, 2011c) and, in context with western Atlantic marine biogeographical patterns, by Petuch (2013). All of these texts together provide a systematic framework for the biodiversity sections of the chapters that follow. The senior author has also described and illustrated 450 new species of fossil gastropods and bivalves from the rich Plio-Pleistocene shell beds of the Everglades region (see Petuch, 1994, 2004; Petuch and Drolshagen, 2011). These beautifully preserved fossils,

Figure 1.1 View of Missouri Key, Middle Florida Keys, showing the complex interfingering of three different marine biotopes: the Vegetated Sedimentary Shore Macrohabitat (mangrove forests, with the *Melampus coffeus* Assemblage and the *Crassostrea rhizophorae* Assemblage); the Unvegetated Rocky Shore Macrohabitat (exposed rocks and tide pools, with the *Cenchritis muricatus* Assemblage, the *Nerita versicolor* Assemblage, and the *Cerithium lutosum* Assemblage); and the Vegetated Softbottom Macrohabitat (adjacent sea grass beds, with the *Modulus calusa* Assemblage). (Photograph by Eddie Matchett.)

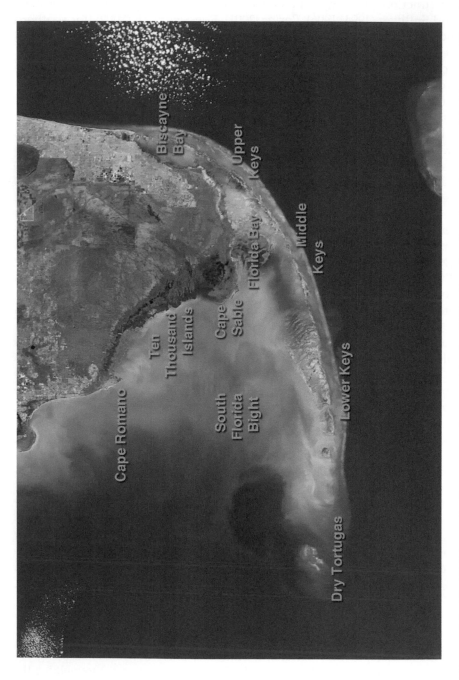

Figure 1.2 Image of the Florida Keys and adjacent areas from an altitude of 100 miles, showing the main geographical features. These include the island chains of the Upper Florida Keys, Middle Florida Keys, Lower Florida Keys, Ten Thousand Islands, and Dry Tortugas; the headlands of Cape Romano and Cape Sable; and the semienclosed embayments of Florida Bay and Biscayne Bay. The Florida Keys Reef Tract and the intervening Hawk Channel parallel the seaward edges of the Upper, Middle, and Lower Florida Keys. The wide, shallow embayment off the Gulf of Mexico, extending from Cape Romano and the Dry Tortugas in the west and connecting to Florida Bay in the east, and encompassing the Florida Keys and Ten Thousand Islands, is here given a new geographical designation, the South Florida Bight. (Base satellite imagery courtesy of NASA.)

several of which are illustrated further in this chapter, represent the precursors of the modern Florida Keys molluscan fauna, and their discovery gives insight into the patterns of evolution and extinction in Neogene southern Florida.

Molluscan faunas of the ancestral Florida Keys

The surface landforms and the underlying geology of the Recent Florida Keys directly control the types of marine habitats that have become established over the past 10,000 years. The present-day geomorphologic features are built on, and are a temporal extension of, the immense coral reef tracts that flourished in southern Florida during the Pliocene and early Pleistocene Epochs. These linear coralline structures were part of a huge atoll-like reef system, referred to as the Everglades Pseudoatoll, which extended northward to Palm Beach County on the east and Lee County on the west (Petuch, 2004; Petuch and Roberts, 2007; Petuch and Sargent, 2011c). This U-shaped reef system enclosed an immense lagoon that was named the Okeechobean Sea (Petuch, 2004: 5–7, 2008; Petuch and Roberts, 2007), and these two marine macroenvironments together laid down the carbonate sedimentation that created the geologic framework for the modern Florida Keys. The Everglades Pseudoatoll itself was built on a depressional feature that formed on the Florida Platform during the late Eocene Epoch, and the present-day geomorphology follows a template established nearly 36 million years ago. Since that time, the Okeechobean Sea has gradually been filling with carbonate sediments, ultimately culminating in the formation of the present-day Everglades (see Petuch and Roberts, 2007: 7–25 for details on the formation of the basin and the effects, in Florida, of the late Eocene Chesapeake Bay asteroid impact and its megatsunamis).

During the early Pleistocene (Aftonian Interglacial Stage of the Calabrian Age, 1.5 million years BP), deposition of carbonates on the Everglades Pseudoatoll reef system had built up to the point that extensive island chains had formed along its southeastern edge (Figure 1.3). Termed the Miami Archipelago (Petuch, 2004: 229, 236), this group of large, wide islands now forms the core of the Atlantic Coastal Ridge, the topographically high geomorphologic feature that underlies the metropolitan areas of modern Miami, Fort Lauderdale, and Palm Beach and the eastern section of Everglades National Park. In response to higher sea levels and warmer water conditions during the Aftonian Stage, a new reef tract began to form along the southern edge of the Miami Archipelago. This fast-growing, zonated reef complex, termed the Monroe Reef Tract (Petuch, 2004: 229; Petuch and Roberts, 2007: 149), now forms the core of the modern Florida Keys island chain, and its remnants are present around 60–70 m below the younger coral rock that is exposed on the surfaces of the central and northern Keys (Petuch and Roberts, 2007: 193). The main platform of the Monroe Reef Tract was dominated by massive heads of the star corals *Montastrea annularis*, *Solenastrea hyades*, and *Siderastrea siderea* and closely resembled the reef platforms seen on the Recent Great Barrier Reef of Belize. Smaller coral species also grew in the more sheltered areas of the main reef platform, including the mushroom corals *Scolymia lacera* and *Mussa angulosa* and the extinct *Thysanus* (unnamed species). The back reef areas of the Monroe Reef Tract contained a rich fauna of more delicate corals, often housing dense thickets of the branching corals *Porites furcata*, *Oculina diffusa*, and the extinct *Arcohelia limonensis* (see Petuch, 2004: 233–238, for illustrations of early Pleistocene corals and descriptions of the early Pleistocene back reef communities).

The coral platform and coral thickets of the Monroe Reef Tract, along with the coral bioherms and shallow carbonate banks of the adjacent Okeechobean Sea, housed one of the richest and most interesting molluscan faunas known from the early Pleistocene tropical

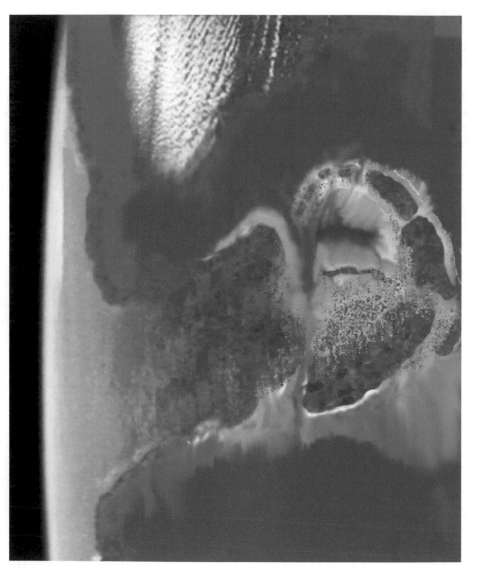

Figure 1.3 Simulated image of Florida as it may have appeared 1.5 million years ago during the Aftonian Interglacial Stage of the early Pleistocene Epoch. From an altitude of 200 miles, the Okeechobean Sea is visible in the bottom half of the image, as is the Miami Archipelago (the chain of wide, vegetated islands) and the ancestor of the modern Florida Keys, the Monroe Reef Tract. (Taken from Petuch, E.J. and C.E. Roberts. 2007. *The Geology of the Everglades and Adjacent Areas.* CRC Press, Boca Raton, FL.)

Americas. This uniquely Floridian prehistoric molluscan fauna is represented by beauti-fully preserved shells that are contained in the Bermont Formation, the geologic unit that encompasses both the Monroe Reef Tract coralline facies and the lime mud (calcarenite) beds that underlie the entire Everglades area. Some of these reef-associated mollusks, such as the cowries *Luria voleki* (Figure 1.4A), *Macrocypraea (Lorenzicypraea) spengleri* (Figure 1.4G, H), and *Macrocypraea joanneae* (Figure 1.4I); the egg shells *Cyphoma tomeui* (Figure 1.4D) and *Cyphoma voleki* (Figure 1.4L); the triviids *Pusula lindajoyceae* (Figure 1.4E) and *Pusula bermontiana* (Figure 1.4F); the cone shells *Lindaconus swearingeni* (Figure 1.5H), *Gradiconus capelettii* (Figure 1.5F), and *Gradiconus holeylandicus* (Figure 1.5G); the muricid *Vokesimurex bellegladeensis* (Figure 1.5B, C); the strombid *Strombus erici* (Figure 1.4K); and the volutid *Scaphella capelettii* (Figure 1.5J), all have descendants that are now living in the Florida Keys. Other typical Monroe Reef Tract mollusks, such as the tonnid *Mulea (Malea) petiti* (Figure 1.4J), the cowrie *Pseudozonaria portelli* (Figure 1.4B), the columbellid *Microcithara caloosahatcheensis* (Figure 1.5L), and the eocypraeid *Jenneria loxahatchiensis* (1.4C), belong to genera that are now extinct in the Atlantic Ocean but are still living along the western side of Central America and northern South America (in the Panamic Molluscan Province). Still others, such as the spectacular giant red-spotted olivid *Lindoliva spengleri* (Figure 1.5A; with specimens of over 130 mm in length, the largest known olivid species) and the large potamidid *Pyrazisinus palmbeachensis* (Figure 1.5D), belong to endemic genera that are only known from the Okeechobean Sea area and are now extinct (other Bermont fossils are shown here in Figures 1.4 and 1.5; see Petuch, 1994, 2004; Petuch and Roberts, 2007; and Petuch and Drolshagen, 2011, for illustrations of the rich Bermont Formation molluscan fauna). The extreme cold climate and lowered sea levels during the subsequent Illinoian Glacial Stage at the end of the Calabrian Age (600,000 years BP) destroyed the Monroe Reef Tract and decimated its unique molluscan fauna.

By the end of the late Pleistocene (Sangamonian Interglacial Stage of the Tarantian Age; 150,000–75,000 years BP), sea levels again began to rise, flooding the Okeechobean Sea area with warm, tropical water. At this time, the Okeechobean Sea had become filled with carbonate sediments, and the main coral reef growth was then concentrated in an area along the southern and southeastern edge of the rapidly growing peninsula (Figure 1.6). This new late Pleistocene reef tract, which grew on top of the remnants of the older early Pleistocene Monroe Reef Tract, is referred to as the Key Largo Reef System (Petuch and Roberts, 2007: 174–175). The eroded remnants of this late Pleistocene carbonate system now form the surficial limestones and coral limestone surfaces of the Upper and Middle Florida Keys (the Key Largo Formation). The coral reef tract that is currently growing offshore of the Florida Keys represents a temporal extension of this late Pleistocene reef system. Another late Pleistocene geological feature that continues to influence the geomorphology of the Florida Keys area is the immense oölitic limestone bank that extends from southern Broward County, across Dade and Monroe Counties, under all of Florida Bay, and west to Boca Grande Key near Key West. Referred to as the Miami Oölite Banks (Petuch and Roberts, 2007: 174–175), this extensive carbonate feature was developed by the massive accumulation of oöids (limestone pellets) that had formed in the heavy surf zone areas of the Key Largo Reef System. These late Pleistocene carbonate pellets washed into the quiet protected areas behind the reefs and produced huge oölite banks that spread across most of then-submerged southern Florida. During sea level drops in the latest Pleistocene (Wisconsinan Glacial Stage), these oölite banks were exposed to subaerial conditions and cemented into an oölitic limestone. Today, this limestone, the Miami Formation, forms the framework for the island chain that extends

from Scout Key and Big Pine Key westward to Key West and Boca Grande Key (Lower Florida Keys) and for the southern section of the Atlantic Coastal Ridge.

By the time of the warm Sangamonian Interglacial Stage, much of the fauna of the older Bermont Formation (both Holey Land and Belle Glade Members) and the Monroe Reef Tract were now extinct or extirpated and were replaced by a much more impoverished set of molluscan assemblages (Petuch and Roberts, 2007; Petuch, 2008). This last late Pleistocene malacofauna, represented by the fossils of the Fort Thompson Formation, Anastasia Formation, and contemporaneous Miami Formation, closely resembled the modern Florida Keys fauna but contained a number of distinct differences. Classic southern Floridian reef-associated taxa such as the mitrid *Nebularia barbadensis* (Figure 1.7A), the triton *Cymatium (Septa) krebsii* (Figure 1.7B), the cowrie *Macrocypraea (Lorenzicypraea) cervus* (Figure 1.7C, D), the strombid *Eustrombus gigas* (Figure 1.7I), and the vase shell *Vasum muricatum* (Figure 1.7E) were all present on the Key Largo Reef System and adjacent areas by this time. Unlike the Recent Florida Keys, however, several species that are now restricted to western Florida and the northern Gulf of Mexico were also present on the reefs and carbonate environments of southeastern Florida. Two of the more prominent of these were the buccinid *Solenosteira cancellarius* (Figure 1.7H; now ranging from northwestern Florida to Texas) and the muricid *Vokesinotus perrugatus* (Figure 1.7L; now restricted to western Florida), both of which lived along with the last-living member of the Okeechobean Sea endemic potamidid genus *Pyrazisinus* (*P. ultimus*, Figure 1.7J). By the end of the late Pleistocene (Wisconsinan Glacial Stage of the Tarantian Age; 75,000–10,000 years BP and the last major sea level drop), the composition of this transitional fauna was completely altered, with the genus *Pyrazisinus* becoming extinct and with the impoverished local fauna having been enriched by an influx of new Caribbean Province migrants (other representatives of this transitional late Pleistocene fauna are shown in Figure 1.7). By the Holocene (11,000 years BP and with rising sea levels flooding previously dry areas), a new modern faunal composition was fully established in the Florida Keys and adjacent areas, and its ecological components are examined in detail in the following chapters of this book.

Figure 1.4 Gastropods of the Monroe Reef Tract and early Pleistocene Okeechobean Sea (Calabrian Age). A = *Luria voleki* Petuch, 2004, length 37 mm (ancestor of the Recent *Luria cinerea*). B = *Pseudozonaria portelli* (Petuch, 1990), length 23 mm (genus extinct in the Atlantic Ocean; extant in the tropical Eastern Pacific). C = *Jenneria loxahatchiensis* M. Smith, 1936, length 29 mm (genus extinct in the Atlantic Ocean; extant in the tropical Eastern Pacific). D = *Cyphoma tomeui* Petuch and Drolshagen, 2011, length 24 mm (ancestor of the Recent *Cyphoma mcgintyi*). E = *Pusula lindajoyceae* Petuch, 1994, length 22 mm (ancestor of the Recent Palm Beach and northern Florida Keys endemic *Pusula juyingae*). F = *Pusula bermontiana* (Petuch, 1994), length 21 mm (related to the Recent *Pusula pediculus*). G, H = *Macrocypraea (Lorenzicypraea) spengleri* (Petuch, 1990), length 121 mm [ancestor of the late Pleistocene and Recent *Macrocypraea (Lorenzicypraea) cervus*]. I = *Macrocypraea joanneae* Petuch, 2004, length 65 mm (ancestor of the Recent *Macrocypraea zebra*). J = *Malea (Malea) petiti* Petuch, 1989, length 122 mm (genus extinct in the Atlantic Ocean; extant in the tropical Eastern Pacific). K = *Strombus erici* Petuch, 1994, length 94 mm (ancestor of the Recent *Strombus alatus*). L = *Cyphoma voleki* Petuch and Drolshagen, 2011, length 22 mm (ancestor of the Recent Florida Keys endemic *Cyphoma rhomba*). All of these fossil gastropods were collected in the lower beds (Holey Land Member; Aftonian Interglacial Stage of the Calabrian Age, early Pleistocene) of the Bermont Formation in Palm Beach and Dade Counties, Florida.

Geomorphology of the Florida Keys, Florida Bay, and the Ten Thousand Islands

The southernmost area of the Florida Peninsula is geologically the youngest and is dominated by a single large feature, the Florida Keys. Dating from the Sangamonian Interglacial Stage of the Tarantian Age of the late Pleistocene, the Keys archipelago consists of four main sections, each classified by its lithologic composition. These are the High Coral Keys (extending from Virginia Key and Key Biscayne to Plantation Key), the Low Coral Keys (extending from Windley Key to Bahia Honda Key), the Oölite Keys (extending from Scout Key and Big Pine Key to Boca Grande Key and Key West), and the Distal Atolls (including the Marquesas Keys and Dry Tortugas) (White, 1970; Petuch and Roberts, 2007: 6–7). Both the High and Low Coral Keys are formed around eroded remnants of the reef platform and back-reef sections of the Key Largo Reef System (Hoffmeister and Multer, 1968; Petuch and Roberts, 2007: 192–193). The Oölite Keys together represent an extension of the oölitic Miami Formation, while the Distal Atolls are new, late Pleistocene-Holocene features that are composed primarily of living coral reefs and cemented coral rubble. The circular atoll-like structure of the Marquesas Keys is unusual, and its origin is problematical. The Florida Keys Reef Tract, which parallels the High and Low Coral and Oölite Keys along their seaward sides, is separated from the islands by the wide Hawk Channel. This shallow (1- to 10-m depths), elongated, depressional feature contains the majority of the open carbonate sand seafloors found in the Florida Keys.

The area stretching from Cape Romano and Cape Sable, along the southwesternmost coast of Florida, encompasses the Reticulated Coastal Swamps (Figure 1.8). Composed primarily of Red and Black Mangrove islands, this estuarine coastline is contiguous with the Big Cypress Swamp area of the Everglades National Park and receives a large input of freshwater. Named for its checkered appearance when seen on aerial photographs, the Reticulated Coastal Swamps contain a series of marine environments that are the ecological intermediates between the quartz sand and seasonally cooler-water macrohabitats of western Florida and the carbonate mud and warmer-water macrohabitats of the Florida

Figure 1.5 Gastropods of the Monroe Reef Tract and early Pleistocene Okeechobean Sea (Calabrian Age). A = *Lindoliva spengleri* Petuch, 1988, length 109 mm (genus extinct; known only from the Bermont Formation fossil beds of the Everglades region and the largest-known olivid). B, C = *Vokesimurex bellegladeensis* (E. Vokes, 1963), length 46 mm (ancestor of the Recent *Vokesimurex morrisoni;* see Figure 8.5K, L in Chapter 8). D = *Pyrazisinus palmbeachensis* Petuch, 1994, length 62 mm (genus extinct; known only from the fossil beds of southern Florida). E = *Cinctura capelettii* (Petuch, 1994), length 78 mm (ancestor of the Recent Dry Tortugas endemic *Cinctura tortugana*). F = *Gradiconus capelettii* (Petuch, 1990), length 36 mm (ancestor of the Recent Florida Keys endemic *Gradiconus mazzolii*). G = *Gradiconus holeylandicus* Petuch and Drolshagen, 2011, length 51 mm (ancestor of the late Pleistocene and Recent Florida Keys endemic *Gradiconus burryae*). H = *Lindaconus swearingeni* Petuch and Drolshagen, 2011, length 54 mm (ancestor of the late Pleistocene and Recent *Lindaconus atlanticus*). I = *Americoliva edwardsae* (Olsson, 1967), length 42 mm (ancestor of the Recent *Americoliva bollingi*). J = *Scaphella capelettii* Petuch, 1994, length 105 mm (ancestor of the Recent *Scaphella junonia;* this fossil species has a much lower spire than does its living descendant). K = *Cariboconus griffini* (Petuch, 1990), length 18 mm (genus extinct in Florida; extant in the Bahamas and central Caribbean). L = *Microcithara caloosahatcheensis* Petuch, 1994, length 18 mm (genus extinct in the Atlantic Ocean; extant in the tropical Eastern Pacific). All of these fossil gastropods were collected in the lower beds (Holey Land Member; Aftonian Interglacial Stage of the Calabrian Age of the early Pleistocene) of the Bermont Formation in Palm Beach and Dade Counties, Florida.

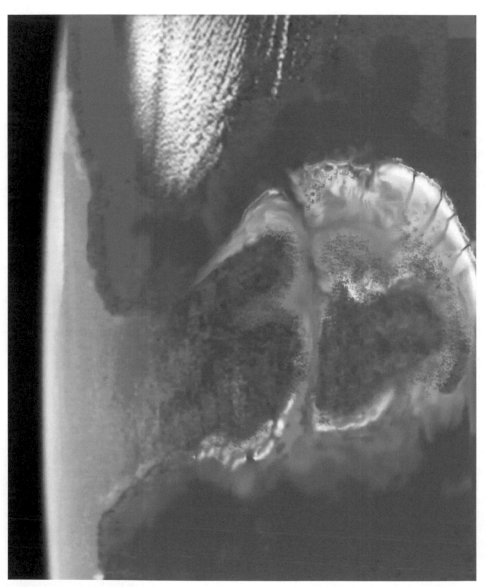

Figure 1.6 Simulated image of Florida as it may have appeared 150,000 years ago during the Sangamonian Interglacial Stage of the late Pleistocene. From an altitude of 200 miles, the almost completely filled Okeechobean Sea is visible in the bottom half of the image, as are the Miami Oölite Banks (white areas) and the ancestor of the modern Florida Keys, the Key Largo Reef Tract (which is built on top of the older early Pleistocene Monroe Reef Tract). (Taken from Petuch, E.J. and C.E. Roberts. 2007. *The Geology of the Everglades and Adjacent Areas.* CRC Press, Boca Raton, FL.)

Keys. Although having relatively impoverished molluscan faunas, the novel ecosystems of the Reticulated Coastal Swamps are now known to contain at least two endemic subspecies. These are discussed and listed in Chapter 7.

The seaward edge of the Reticulated Coastal Swamps houses the Ten Thousand Islands, a unique archipelago that is actually made up of a labyrinth of over 11,000 low, tightly packed keys and is composed of a bizarre mixture of worm gastropod reefs, oyster banks, and Red and Black Mangrove jungles. Separating the Ten Thousand Islands from the Big Cypress paludal coastline is a series of large open-water brackish lakes, referred to as the Back Bays (White, 1970: 43). Some of these, such as Chokoloskee Bay, retain a fairly high salinity all year round, while others can have low salinities during the rainy summer months. The Ten Thousand Islands themselves fall into two broad categories: the Outer Ten Thousand Islands (distal seawardmost), which are built on unusual massive vermetid gastropod bioherms ("worm shell reefs"; discussed in Chapter 7), and the Inner Ten Thousand Islands (proximal landwardmost), which are composed primarily of mangrove peat, mud, and oyster shells (*Crassostrea virginica*, *Crassostrea rhizophorae*, and *Ostreola equestris*). No similar worm shell reef, oyster bank, and mangrove forest amalgam is known from anywhere else in the western Atlantic.

Between Cape Sable and the High and Low Coral Keys (Upper and Central Keys) lies the shallow, funnel-shaped embayment of Florida Bay. Connecting to Blackwater Sound in the northeast, Florida Bay widens to the west and southwest, opening into a broad embayment off the Gulf of Mexico north of the Lower Keys (Oölite Keys). Throughout its extent, Florida Bay is composed of a densely intertwined network of narrow, shallow banks, mangrove islands, and vast expanses of Turtle Grass (*Thalassia testudinum*). This distinctive network of narrow banks is obvious even on photographic images from satellites. Most of the Turtle Grass-based molluscan communities found in the Florida Keys area are located within Florida Bay. This sea grass-dominated area characteristically houses a large number of endemic mollusks, making Florida Bay one of the most important centers for speciation occurring anywhere along eastern North America. The Turtle Grass environments of Florida Bay also house the richest sea grass-based molluscan fauna found in the entire western Atlantic (see Chapter 4; also Petuch, 2013: 48–50). The large, shallow embayment off the Gulf of Mexico, extending from the Dry Tortugas and Cape Romano in the west and connecting to Florida Bay in the east and encompassing the Florida Keys and the Ten Thousand Islands, is referred to as the South Florida Bight, a new geographical feature proposed here for the first time. With the exception of the coral coasts of Dade, Broward, and Palm Beach Counties to the northeast, all the environments and molluscan assemblages discussed in this book are unique to the area of the South Florida Bight.

The northern sides of the westernmost Oölite Keys and the Distal Atolls slope rapidly downward to form a wide and deep (100- to 200-m depths) platform. These areas, particularly those north of the Dry Tortugas along the mouth of the South Florida Bight, are under the influence of upwelling systems and contain high-productivity water conditions. Because of this, the extreme distal region is a major shrimping and lobstering area and has been commercially fished for the past century. Sedimentological data has shown that this deep neritic terrace contains huge expanses of carbonate sand and coralline algae beds, each containing a different set of molluscan assemblages. Off the southern and eastern sides of the High and Low Coral Keys and the southern side of the Oölite Keys, the narrow Florida Reef Tract drops off precipitously in a series of step-like terraces, ranging from 100- to 400-m depths. Unlike the deep neritic terrace north of the Distal Atolls at the mouth of the South Florida Bight, these terraces are composed of both talus slopes of coral debris that had tumbled down from the steep reef faces and narrow shelves covered with fine

carbonate mud and pelagic sediments. Some of the rarest and least-known mollusks reside within the communities found on these deep terraces; these are discussed in Chapters 8 and 9.

Marine ecological units of the Florida Keys and adjacent areas

A standardized hierarchical classification scheme for the marine ecosystems of the South Florida Bight, Florida Keys, and Florida Bay was not formulated until 2007, when Madden and Goodwin first proposed their CMECS (Coastal Marine Ecological Classification Standard) system. Until that time, most ecological studies involved individualized inter-pretations that were skewed toward the specialty organisms of each author, such as sea grasses, coral reefs, or mangroves (as discussed in the first sections of this chapter). The eco-logical classification scheme proposed by Madden and Goodwin was the first to incorporate aspects of bathymetry, substrate type (biotopes), geomorphology, and organismal biology. Because of its wide-reaching and holistic approach, the CMECS model has been adopted by most government agencies and private-sector environmental survey companies. When the molluscan ecological data accumulated for this book was applied to the CMECS scheme, an almost-perfect categorical fit presented itself. Every molluscan fauna and molluscan-based biotope known from the South Florida Bight, Florida Keys, Florida Bay, the Ten Thousand Islands, and adjacent deep-water areas corresponds to one of the CMECS units, producing a hierarchical arrangement of ecosystems that allows for higher-order analyses of ecological relationships. For this reason, we have adopted Madden and Goodwin's CMECS hierarchy for the molluscan community classification outlined in this book.

The CMECS for the South Florida Bight, Florida Keys, and Florida Bay takes into account three levels of bathymetrically controlled biotopes that, in turn, influence a sin-gle set of biotic units. The highest biotope level is that of the oceanic regime, which in turn is composed of individual geomorphological formations (structural biotopes). Each

Figure 1.7 Gastropods of the Key Largo Reef Tract and adjacent areas (late Pleistocene). A = *Nebularia barbadensis* (Gmelin, 1791), length 25 mm (extant in the Recent Florida Keys). B = *Cymatium (Septa) krebsii* Mörch, 1877, length 28 mm (extant in the Recent Florida Keys). C, D = *Macrocypraea (Lorenzicypraea) cervus* (Linnaeus, 1771), length 119 mm (extant in the Recent Florida Keys). E = *Vasum muricatum* (Born, 1778), length 84 mm (extant in the Recent Florida Keys). F = *Gradiconus burryae* (Clench, 1942), length 38 mm (this specimen was incorrectly identified as *G. patglicksteinae* in Petuch, 2004, and Petuch and Roberts, 2007; the species is extant in the Recent Florida Keys). G = *Jaspidiconus pfluegeri* Petuch, 2004, length 21 mm (extant in the Recent Florida Keys and Palm Beach coast). H = *Solenosteira cancellarius* (Conrad, 1846), length 29 mm (extinct in southeastern Florida; extant along northwestern Florida and along the northern Gulf of Mexico as far as Texas). I = *Eustrombus gigas* (Linnaeus, 1758), length 178 mm (also found in the con-temporaneous late Pleistocene Fort Thompson, Miami, and Anastasia Formations; extant in the Florida Keys). J = *Pyrazisinus ultimus* Petuch, 2004, length 52 mm (genus now extinct, and this is the last-living species; *Pyrazisinus* ranges from the late Oligocene to the latest Pleistocene). K = *Lindaconus atlanticus* (Clench, 1942), length 55 mm (extant in the Recent Florida Keys). L = *Vokesinotus perrugatus* (Conrad, 1836), length 22 mm (extinct in southeastern Florida but extant along western Florida; this is the last-living member of the eastern American Neogene muricid genus *Vokesinotus*, which reached its evolutionary peak during the Pliocene. Often incorrectly placed in the genus *Urosalpinx*.) All of these fossil gastropods were collected in the upper beds (Coffee Mill Hammock Member; Sangamonian Interglacial Stage of the Tarantian Age) of the Fort Thompson Formation in Palm Beach and Dade Counties, Florida.

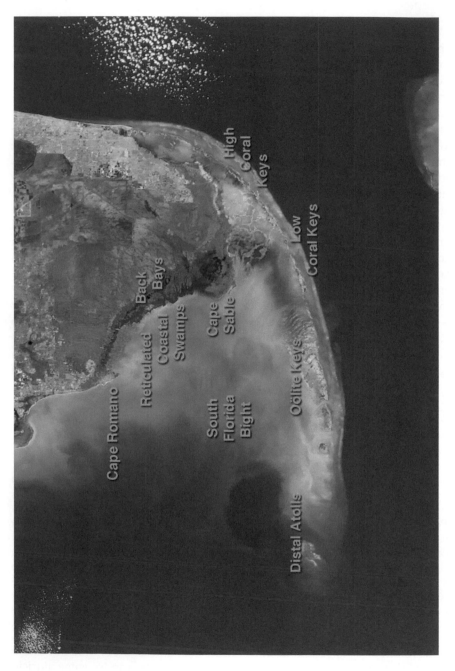

Figure 1.8 Image of the Florida Keys and adjacent areas from an altitude of 100 miles, showing the main geomorphological structures. These include island archipelagoes such as the High Coral Keys and Low Coral Keys (built on the coral limestone of the Key Largo Formation), the Oölite Keys (built on the oölitic limestone of the Miami Formation), and the Distal Atolls (composed of Recent coral reefs and coral rubble); estuarine and paludal areas such as the Reticulated Coastal Swamps (composed of vermetid worm shell "reefs," oyster banks, and mangrove forests) and Back Bays (brackish water lagoons in mangrove forests); and headlands such as Cape Romano and Cape Sable (composed of quartz sand transported by longshore currents from western Florida). All of these landforms are contained within the South Florida Bight and Florida Bay. (Base satellite imagery courtesy of NASA.)

formation consequently contains sets of ecological macrohabitat biotopes that, individually, support an array of molluscan assemblages (biotic units). The boundaries of the macrohabitats listed here generally are not sharply defined but often blur together in a series of ecotonal transition zones. The regimes, formations, and macrohabitats for the region covered by this book (originally proposed by Madden and Goodwin) include the following:

BIOTOPE-INFLUENCED UNITS
    REGIMES (reflecting overall oceanographic parameters)
        Estuarine (brackish water areas, shallow coastal lagoons, mangrove forests)
        Nearshore Marine (sea grass beds, mud flats; coral reefs; sponge bioherms; deep lagoons)
        Neritic (deep coral reefs, deep reef talus slopes)
        Oceanic (deep offshore terraces, deep offshore coralline algal beds, pleuston)
    FORMATIONS (reflecting geomorphological patterns)
        Land (hypersaline pools, terrestrial interface)
        Lagoon (mangrove forests, sea grass beds, open sand seafloors)
        Shallow Coral Reef (zonated coral reefs, coral bioherms)
        Noncoral Reef (vermetid gastropod reefs)
        Deep Coral Reef (deep photic zone hermatypic corals)
    MACROHABITATS (reflecting individual biotopes)
        Vegetated Sediment Shore
        Unvegetated Rocky Shore
        Vegetated Softbottom
        Unvegetated Hardbottom
        Unvegetated Softbottom
        Linear Coral Reef
        Spur and Groove Platform Reef
        Individual Patch Reef
        Aggregated Patch Reef
        Gastropod Reef
        Deep Reef Talus
        Deep Softbottom
        Pelagic

The Vegetated Sediment Shore Macrohabitat encompasses forests of the Red Mangrove (*Rhizophora mangle*), Black Mangrove (*Avicennia germinans*), and White Mangrove (*Laguncularia racemosa*) and thickets of the Saltwort (*Salicornia* spp.), all of which grow on unconsolidated mud and sand. Within the CMECS scheme, this macrohabitat falls under the Estuarine Regime and Lagoon Formation and grades directly into the terrestrial Land Formation. This transition is facilitated by a series of vegetative interfaces, especially those provided by ecotonal plants such as Beach Morning Glories, Sea Grapes, and grasses. The salt-tolerant land snail genus *Cerion* (with four endemic subspecies in the Florida Keys) bridges this terrestrial-marine gap and is an indicator of this type of ecotonal habitat (see Chapter 2).

In the Florida Keys, the Unvegetated Rocky Shore Macrohabitat includes the exposed "iron rock" (eroded limestone) shorelines and tide pools of supratidal and intertidal limestone platforms. Characteristically, these rocky areas are devoid of both angiosperms and macroalgae and are exposed to extremes in temperature and salinity. A richer intertidal rocky shore malacofauna is present on the exposed outcrops of the Anastasia Formation found along the Palm Beach coast farther north. Here, the low cliffs and heavier wave action create more available microhabitats and a broader vertical area for habitation. Because of these factors, the Palm Beach rocky coastline houses a number of endemic species (see Chapter 6). The Florida Keys tide pools, found along the seaward sides of both the coral and oölite islands, also house several endemic mollusks, primarily the dwarf melongenid gastropod *Melongena* (*Rexmela*) *bicolor*. Within the CMECS scheme, this macrohabitat falls under the Nearshore Marine Regime and Land Formation.

The Vegetated Softbottom Macrohabitat falls within the Nearshore Marine Regime and the Lagoon Formation and consists of all intertidal-to-subtidal sand or mud substrates that support dense growths of sea grasses and macroalgae. Within the South Floria Bight, Florida Keys, and Florida Bay area, the principal sea grasses include Turtle Grass (*Thalassia testudinum*), Manatee Grass (*Syringodium filiforme*), and Shoal Grass (*Halodule wrightii*); these generally segregate themselves by depth, with *Thalassia* and *Syringodium* occurring in the deepest water (averaging 1- to 10-m depths) and *Halodule* occurring in the shallowest intertidal conditions. In some areas, such as on the labyrinthine shoals of Florida Bay and in Lake Worth Lagoon, transition zones occur where two or all three of these species grow together sympatrically. Because of its large leaf size, closely packed growth habit, and dense rhizome root mat, *Thalassia* provides the ideal microtopography for supporting the richest grass-associated invertebrate community. In the Florida Keys, the mollusk and echinoderm faunas of the Turtle Grass beds are second in species richness only to those of the coral reef platforms.

Within the Florida Keys area, the Unvegetated Hardbottom Macrohabitat is confined to the Nearshore Marine Regime and Lagoon Formation and is generally highly localized in distribution. Composed of oölitic limestone ramps (Miami Formation) that are covered by thin (1- to 4-cm) layers of clean carbonate sand, this macrohabitat is best developed in the channels between the Oölite Keys and in shallow areas bordering the northern sides of the Oölite Keys and the Distal Atolls. Here, wide expanses of eroded microkarstic limestone occur just below a thin sand veneer; these provide the substrate for the attachment of a rich fauna of sessile invertebrates. Some of the more prominent types include solitary zooantharian cnidarians and large sponges, with the latter often forming dense biohermal structures (Petuch and Sargent, 2011a). These "sponge reefs," which are particularly well developed along the northern sides of the Lower Keys and Distal Atolls, house a rich and highly endemic molluscan fauna and are still virtually unstudied and unexplored. Other variants of the Unvegetated Hardbottom Macrohabitat include beds of the Yellow Mussel *Brachidontes modiolus*; thickets of the shallow-water gorgonian octocorals *Antillogorgia*, *Plexaurella*, *Eunicea*, and *Pterogorgia*; and open ramps of highly eroded coral limestone (Key Largo Formation) dominated by the rock-boring sea urchin *Echinometra lucunter* (all discussed in Chapter 5). Like the sponges and sea anemones, the mussels and gorgonians attach themselves to the hard limestone substrate and extend above the thin covering of carbonate sand.

Along the Florida Keys, throughout Florida Bay, and northward to the Palm Beach coast, the most bathymetrically expansive set of biotopes is that of the Unvegetated Softbottom Macrohabitat. Extending from the high intertidal zone to deeper subtidal areas in channels (down to 50-m depth), this macrohabitat encompasses open mud and sand flats, open sand seafloors, and tidal creeks in coastal lagoons. These open softbottom areas, particularly the clean carbonate sand seafloor in the Hawk Channel, support a rich fauna of sand-dwelling and burrowing echinoid echinoderms. Some shallow-burrowing forms, such as the sand dollar *Mellita*, are the principal prey of the small cassid gastropods (Scotch Bonnets) *Semicassis granulata*, *Semicassis cicatricosa*, and *Casmaria atlantica*. Other echinoids, such as the large heart urchins *Meoma* and *Plagiobrissus*, serve as the main prey items of the giant cassids (Helmet Shells) *Cassis spinella*, *Cassis madagascariensis*, and *Cassis tuberosa*. All of these cassid species, along with two closely adjacent coral reef and rubble-dwelling species (*Cassis flammea* and *Cypraecassis testiculus*) combine to produce the single largest sympatric fauna of Cassidae found anywhere on Earth. This unusually rich shallow-water evolutionary radiation, composed of eight species in four genera, is unique to the Unvegetated Softbottom Macrohabitat of the Florida Keys (see Chapter 6).

Based on patterns of geomorphology and growth forms, Madden and Goodwin (2007) divided the living coral reef areas of the Florida Reef Tract into four separate ecological types: the Linear Coral Reef Macrohabitat, the Spur and Groove Platform Reef Macrohabitat, the Individual Patch Reef Macrohabitat, and the Aggregated Patch Reef Macrohabitat. Together, these make up the Shallow Coral Reef Formation of the Nearshore Marine Regime. The Linear Coral Reef Macrohabitat is the most extensive type, making up over 90% of the Florida Keys coral reefs and typified by wide, linearly elongated reef platforms, such as those seen on French, Carysfort, Molasses, and Pickles Reefs off the Upper Keys. The Spur and Groove Platform Reef Macrohabitat occurs along the seaward edges of linear reef complexes that receive almost continuous wave action. This coralline macrohabitat characteristically is composed of alternating elongated shallow coral knolls ("spurs") and deep sand-filled gullies ("grooves"), all oriented perpendicular to the wave action. Found mostly off the Middle and Lower Florida Keys where seasonal trade winds produce heavy wave action, these spur-groove reefs are found primarily on Looe Key (the best developed) and sections of Sombrero and Alligator Reefs and Sand Key.

The other two reefal macrohabitats of Madden and Goodwin, the Individual and Aggregated Patch Reefs, represent nonzonated coral bioherms. These small-scale patch reefs are scattered throughout the South Florida Bight and Florida Bay and form on open hardbottom surfaces that are exposed to the oceanographic conditions needed for coral growth. Because the molluscan fauna of the patch reefs, spur-groove reefs, and platform reefs are essentially the same, all the reef types are combined into a single new molluscan ecological category, the Coral Reef Tract Macrohabitat. This combined coral macrohabitat supports an extremely rich invertebrate fauna, including the most ecologically diverse and species-rich molluscan fauna found in the continental United States. Another massive biohermal structure that is included within the Nearshore Marine Regime is that of the Noncoral Reef Formation and the Gastropod Reef Macrohabitat. Unique to the outermost keys of the Ten Thousand Islands, these large barrier reef-type structures are composed of monocultures of the pseudocolonial vermetid gastropod *Vermetus* (*Thylaeodus*) *nigricans*. These extensive worm shell reefs provide habitats for a wide variety of other mollusks, several of which are endemic to the Ten Thousand Islands. Both the coral and worm gastropod reef structures are discussed in Chapter 7.

The third type of large-scale ecological zone that was recognized by Madden and Goodwin is that of the Neritic Regime, which encompasses the narrow strip of deep water (100- to 300-m depths) along the base of the Florida Coral Reef Tract. This regime includes the Deep Coral Reef Formation and its Deep Reef Talus Macrohabitat, an extensive area of coral rubble and dead coral slabs that have tumbled down the steep seaward slopes of the living coral reefs. This deep-water, high-angle coral rubble slope is cemented by hexactinellid sponges and supports a rich fauna of invertebrates, such as antipatharian gorgonians, comatulid crinoids, and large euspongia poriferans. Because of the difficulty of sampling the biota of the steeply angled Deep Reef Talus Macrohabitat, little is known of the composition of the malacofauna. The distinctive mollusks of the Deep Coral Reef Formation range from the Dry Tortugas all the way northward to the coral reefs off Palm Beach County (see Chapter 8).

The outermost seaward edge of the Florida Reef Tract and Dry Tortugas encompasses the Oceanic Regime. Within this large-scale ecological zone, the Deep Softbottom Macrohabitat predominates and typically occurs on the deep (200- to 400-m depths) narrow terraces that edge the Florida Platform along the Florida Straits or on the deep carbonate ramp west and north of the Dry Tortugas. The deep terraces characteristically are covered with thick layers of carbonate mud, foraminiferal oozes, and pelagic sediments. In contrast, the carbonate ramp north of the Dry Tortugas supports extensive beds of the deep-water red coralline alga *Porolithon* and bioherms of the deep-water purple lettuce coral *Agaricia*. The rhodolith nodules of the coralline algae accumulate in immense aggregations, often several meters thick, and this substrate supports an extremely rich and highly endemic molluscan fauna. The Oceanic Regime also includes the open-ocean areas of the southeastern Gulf of Mexico and the Florida Straits, within the influence of the Gulf Loop and Florida Currents. Extending beyond the continental shelf and deep terraces of the Florida Keys, this area houses the Pelagic Macrohabitat and its pleustonic (surface-floating organisms) and planktonic (drifting organisms) molluscan faunas. These three deep-water and open-oceanic ecological zones are discussed in Chapter 9.

Molluscan assemblages of the Florida Keys, Florida Bay, and the Ten Thousand Islands

Each of Madden and Goodwin's macrohabitats contains one or more characteristic Molluscan Assemblages. These faunal units are composed of the resident mollusks that occur on, and are restricted to, a specific biotope. Some species of mollusks are ecologically plastic and can occur in two or more macrohabitats. A classic eurytopic gastropod of this type is the algal film-feeding batillariid *Batillaria minima*, which lives in several intertidal habitats, including the Unvegetated Rocky Shore Macrohabitat, the Vegetated Softbottom Macrohabitat, and the Unvegetated Softbottom Macrohabitat. Other specialized stenotopic gastropods, such as the ectocommensal pectinoidean-associated amathinid *Amathina pacei*, the coral-feeding muricid *Babelomurex scalariformis*, and the fish-eating cone shell *Chelyconus ermineus*, are tightly restricted to only the Coral Reef Tract Macrohabitat. The vast majority of the Florida Keys and South Florida Bight gastropods and bivalves, however, inhabit one, two, or sometimes three of the 10 macrohabitats that are covered in this book.

Altogether, the macrohabitats recognized here have been found to contain 20 distinct molluscan assemblages. These groupings of sympatric mollusks are named for the most conspicuous and characteristic resident species, and these taxa act as proxies for the entire malacofauna. The molluscan assemblages are listed here by regime and macrohabitat, along with their preferred substrate types. These include the following:

ESTUARINE REGIME
Vegetated Sediment Shore Macrohabitat
    Melampus coffeus Assemblage (Black Mangrove forests)
    Crassostrea rhizophorae Assemblage (Red Mangrove forests)
    Cerithideopsis costatus Assemblage (hypersaline pools–terrestrial interface)
NEARSHORE MARINE REGIME
Unvegetated Rocky Shore Macrohabitat
    Cenchritis muricatus Assemblage (supratidal rocky outcrops)
    Nerita versicolor Assemblage (intertidal rocky outcrops)
    Cerithium lutosum Assemblage (subtidal rocky outcrops)

Vegetated Softbottom Macrohabitat
Bulla occidentalis Assemblage (intertidal Shoal Grass beds)
Modulus calusa Assemblage (Turtle Grass beds)
Unvegetated Hardbottom Macrohabitat
Cerodrillia clappi Assemblage (sponge bioherms)
Cyphoma rhomba Assemblage (gorgonian thickets)
Bayericerithium litteratum Assemblage (sublittoral exposed hardbottoms)
Unvegetated Softbottom Macrohabitat
Batillaria minima Assemblage (open intertidal mud flats)
Polinices lacteus Assemblage (sublittoral open carbonate sand)
Coral Reef Tract Macrohabitat
Stephanoconus regius Assemblage (living coral reefs)
Gastropod Reef Macrohabitat
Vermetus nigricans Assemblage (vermetid gastropod reefs)
NERITIC REGIME
Deep Reef Talus Macrohabitat
Cymatium rehderi Assemblage (deep reef talus slopes)
Arca rachelcarsonae Assemblage (deep reef ledges)
OCEANIC REGIME
Deep Softbottom Macrohabitat
Chicoreus rachelcursonae Assemblage (deep coralline algal beds)
Rehderia schmitti Assemblage (deep softbottom terraces)
Open Oceanic Pleustonic Macrohabitat
Janthina janthina Assemblage (floating on bubble rafts in the open ocean)

Several of these molluscan assemblages and their associated biotopes can occur together along one single stretch of coastline, often interfingering in a complex network of ecotones and transitional sedimentary lithofacies. A classic example of this complex interfingering pattern is seen in the Big Pine–Little Torch Channel in the Lower Keys, where the sponge-based *Cerodrillia clappi* Assemblage and the Turtle Grass-based *Modulus calusa* Assemblage form a "patchwork quilt" arrangement of randomly interspersed ecosystems. A similar interfingering pattern is seen along Missouri Key in the Middle Keys, where the *Modulus calusa* Assemblage, the exposed hardbottom *Bayericerithium litteratum* Assemblage, the gorgonian thickets, and the Yellow Mussel beds are all contiguous. Although occurring in close proximity, these assemblages and macrohabitats retain their own distinctive faunal identities, with few species shared between them. These ecotonal patterns are typical of the entire area of the South Florida Bight, Florida Bay, Biscayne Bay, and the coral coast of Palm Beach. Bathymetric interfingering also occurs throughout these areas, with many of the Nearshore Marine Regime macrohabitats and molluscan assemblages forming ecotones with the deeper water macrohabitats and molluscan assemblages of the Neritic Regime. These bathymetric ecotones are especially prevalent along the narrow Palm Beach continental shelf area, where shallow-water reef communities (such as the *Stephanoconus regius* Assemblage) often blur directly into deeper-water assemblages (such as the *Cymatium rehderi* Assemblage). This pattern of faunal overlap is also seen along the reef tracts of the Florida Keys.

Close-up of a clump of the Mangrove Oyster, *Crassostrea rhizophorae*, growing on the prop roots of a Red Mangrove tree in Lake Worth Lagoon, Palm Beach County, Florida. Mangrove oysters from this sheltered environment are especially beautiful, with large fluted spines and serrated shell margins. (Photograph by Eddie Matchett)

chapter two

Molluscan faunas of the vegetated sediment shore macrohabitat (estuarine regime)

Introduction

In the Florida Keys, the Vegetated Sediment Shore Macrohabitat supports three different plant-based ecosystems: one centered on the Black Mangrove (*Avicennia germinans*) forests, one centered on the Red Mangrove (*Rhizophora mangle*) forests, and one centered on Saltwort (*Salicornia* spp.) and Sea Purslane (*Sesuvium portulacastrum*) thickets growing in and near hypersaline and brackish pools and salt flats. All three of these vegetation types prefer muddy, soft sediment environments and are the first land-producing plants to become established along low-energy sediment-filled shorelines. The pioneer Red Mangrove trees, with their densely intertwined network of prop roots and massive growths of oysters, are iconic and embody the mystery of the dark reaches of the impenetrable coastal jungles. The beauty of these alien forests, half in and half out of seawater, was elegantly described by Rachel Carson when she stated that: "A mangrove forest, its fringing trees literally standing in salt water, extending back into swamps of its own creation, is full of the mysterious beauty of massive and contorted trunks, of tangled roots, and of dark green foliage spreading an almost unbroken canopy. The forest with its associated swamp forms a curious world" (1955, pp. 242–243).

Molluscan ecology of the Black Mangrove forests

Of the two partially submerged mangrove species, the Black Mangrove prefers higher ground, often growing in areas that are flooded only during the highest spring tides. Here, in the quiet low-energy water conditions of the Black Mangrove swamp, rotting leaves and organic matter accumulate in thick layers, producing characteristic peat deposits. These mangrove-derived soils are saturated with hydrogen sulfide, and oxic conditions exist for only a few millimeters below the sediment surface. To facilitate the exchange of vital gases such as oxygen and nitrogen, the mud-dwelling Black Mangrove has evolved a system of fringe-like pneumatophores that grow directly from its shallow subsurface roots (Figure 2.1). These vertically oriented "breather roots" provide areas of attachment for numerous sessile bivalves and create the ideal microhabitat for an unusually rich fauna of air-breathing ellobiid gastropods.

In the higher areas among the dense growths of pneumatophores, the large mytilid bivalve *Geukensia granosissima* often occurs in dense aggregations, with each individual animal attached to a pneumatophore by a network of tough byssal threads (Figures 2.2 and 2.3A). In the lower areas of the Black Mangrove forests, the dense thickets of pneumatophores serve as areas of attachment for three sympatric isognomonid oysters: *Isognomon alatus* (Figure 2.3B), *Isognomon bicolor* (Figure 2.3C), and *Isognomon radiatus* (Figure 2.3D). Like

Figure 2.1 View of a Black Mangrove (*Avicennia germinans*) forest along the shoreline of Lake Worth, Palm Beach County. Forests like this extend all along the Florida coastline, from St. Augustine in eastern Florida to Cedar Key in western Florida, and are best developed in Florida Bay and the Florida Keys. Black Mangrove forests are components of the Vegetated Sediment Shore Macrohabitat and support the *Melampus coffeus* Assemblage. (Photograph by Eddie Matchett.)

Figure 2.2 Close-up of an aggregation of the mangrove mussel, *Geukensia granosissima* (Sowerby III, 1914), attached to mangrove roots along Singer Island, Lake Worth Lagoon, Palm Beach County. Numerous specimens of the eurytopic intertidal batillariid gastropod, *Batillaria minima* (Gmelin, 1791), can be seen crawling on the roots and mangrove peat. (Photograph by Robert Myers.)

Figure 2.3 Mollusks of the *Melampus coffeus* Assemblage. A = *Geukensia granosissima* (Sowerby III, 1914), length 95 mm. B = *Isognomon alatus* (Gmelin, 1791), length 103 mm. C = *Isognomon bicolor* (C.B. Adams, 1845), length 58 mm. D = *Isognomon radiatus* (Anton, 1838), length 92 mm. E, F = *Melampus bidentatus* Say, 1822, length 10 mm. G, H = *Melampus coffeus* (Linnaeus, 1758), length 18 mm. I = *Melampus coffeus* (Linnaeus, 1758), length 16 mm. J, K = *Melampus monilis* (Bruguiere, 1798), length 15 mm.

the *Geukensia* mussels, the three "Tree Oysters" attach themselves to the pneumatophores with strong byssal threads and are often partially buried in the mangrove peat. These four bivalves, three isognomatids and one mytilid, are the largest and most prominent invertebrates within the Black Mangrove forests and, between them, contain the bulk of the molluscan biomass for the entire ecosystem. In the Florida Keys, *Isognomon alatus* is the most abundant of the three tree oysters, and *Isognomon radiatus* is the least frequently encountered.

The three *Isognomon* oysters and the *Geukensia* mussel are subject to heavy predation by raccoons, and their empty shells are often heaped up in piles within the pneumatophore thickets. In the Florida Keys and South Florida Bight area, four separate subspecies of the North American Raccoon (*Procyon lotor*) occur, with each geographically isolated from the others. These major predators on the mangrove forest bivalves include the Matecumbe Key Raccoon (subspecies *inesperatus*), which ranges throughout the Upper Keys from Key Biscayne to Upper Matecumbe Key; the Vaca Key Raccoon (subspecies *auspicatus*), which ranges from Grassy Key to Vaca Key and adjacent smaller keys; the Torch Key Raccoon (subspecies *incautus*), which ranges from No Name Key to Key West; and the Ten Thousand Islands Raccoon (subspecies *marinus*), which ranges throughout the mangrove islands between Cape Romano and Cape Sable. With few other available food resources in the maze of small mangrove keys of the Ten Thousand Islands, *Procyon lotor marinus* feeds almost exclusively on sessile marine bivalves such as the oysters *Crassostrea*, *Ostreola*, and *Isognomon* and the mussel *Geukensia*.

Although not as dominant in overall biomass, the representatives of the gastropod family Ellobiidae alone account for the largest number of species in the Black Mangrove forests. At least 16 species, encompassing 11 genera, are now known from the Florida Keys and adjacent areas, making the Ellobiidae one of the largest gastropod families in Florida. These small air-breathing snails, typified by the Coffee Bean Shells *Melampus coffeus* (Figures 2.3G–I and 2.4), *Melampus monilis* (Figure 2.3J, K), and *Melampus bidentatus* (Figure 2.3E, F), feed on algal and bacterial films and on rotting leaf litter and are gregarious, often occurring in aggregations of thousands of individuals. Large ellobiids, such as *Melampus coffeus* and *Melampus monilis*, are the most obvious species and are the most easily collected; tiny species such as *Apodopsis novimundi* and *Pedipes mirabilis* are far more cryptic and are easily overlooked. On the larger mangrove islands of the Key West National Wildlife Refuge, the three *Melampus* species provide a food resource for the endemic saltwater turtle, *Malaclemys terrapin rhizophorarum* (the Mangrove Terrapin, which grazes on the ellobiids at low tide). Because the air-breathing semiland snail *Melampus coffeus* is the most abundant and characteristic gastropod in the Black Mangrove areas, we have chosen this taxon to represent the entire molluscan assemblage and the molluscan community.

Biodiversity of the Melampus coffeus Assemblage

Because of the harsh and mutable environmental conditions of the Black Mangrove forests, the molluscan fauna of the *Melampus coffeus* Assemblage is relatively impoverished, incorporating only 19 species of macrogastropods and 4 species of macrobivalves. These are listed here by feeding type and ecological niche.

1. HERBIVORES (also plant detritus)
Gastropoda
Batillariidae
Batillaria minima (Gmelin, 1791) (eurytopic; occurs in several macrohabitats)

Figure 2.4 Close-up of two specimens of the ellobiid gastropod, *Melampus morrisoni* Martins, 1996, crawling on the trunk of a Black Mangrove on Missouri Key, Middle Florida Keys. (Photograph by E. J. Petuch.)

Truncatellidae
  Truncatella caribaeensis Reeve, 1842
  Truncatella pulchella Pfeiffer, 1839
Ellobiidae
  Apodopsis novimundi Pilsbry and McGinty, 1949
  Blauneria heteroclita (Montagu, 1808)
  Detracia bulloides (Montagu, 1808)
  Detracia floridana (Pfeiffer, 1856)
  Ellobium (*Auriculoides*) *dominicense* (Ferrusac, 1821) (= *E. pellucens*)
  Laemodonta cubensis (Pfeiffer, 1854)
  Marinula succinea (Pfeiffer, 1854)
  Melampus bidentatus (Say, 1822) (Figure 2.3E, F)
  Melampus coffeus (Linnaeus, 1758) (Figure 2.3G–I)
  Melampus monilis (Bruguiere, 1792) (Figure 2.3J, K)
  Melampus morrisoni (Martins, 1996)(Figure 2.4)
  Microtralia occidentalis (Pfeiffer, 1854)
  Ovatella myosotis (Draparnaud, 1801)
  Pedipes mirabilis (Megerle von Muhlfeld, 1816)
  Pedipes ovalis (C.B. Adams, 1849)
  Tralia ovula (Bruguiere, 1792)
2. **CARNIVORES**
  No molluscan carnivores; (for bivalves) replaced by the Raccoons *Procyon lotor auspicatus*, *Procyon lotor incautus*, *Procyon lotor inesperatus*, and *Procyon lotor marinus* and (for ellobiid gastropods) by the Mangrove Terrapin *Malaclemys terrapin rhizophorarum*)
3. **SUSPENSION/FILTER FEEDERS**
  Bivalvia
  Mytilidae
    Geukensia granosissima (Sowerby III, 1914) (Figure 2.3A)
  Isognomonidae
    Isognomon alatus (Gmelin, 1791) (Figure 2.3B)
    Isognomon bicolor (C.B. Adams, 1845) (Figure 2.3C)
    Isognomon radiatus (Anton, 1838) (Figure 2.3D)

See Abbott (1974) and Warmke and Abbott (1962) for photographs of species not illustrated in this book. A review of the niche preferences of the members of the *Melampus coffeus* Assemblage shows that herbivores dominate the Black Mangrove forests.

Molluscan ecology of the Red Mangrove forests

The Red Mangroves, with their interwoven network of stilt-like prop roots (Figure 2.5), provide the substrate for a rich molluscan-based ecosystem. In the Florida Keys and adjacent areas of the South Florida Bight, the prop roots of the Red Mangroves are typically heavily encrusted with the Mangrove Oyster, *Crassostrea rhizophorae*, and the immense aggregations of these sessile bivalves support a rich and ecologically diverse malacofauna. As the most conspicuous molluscan faunal component of the Red Mangrove community, we have here selected *Crassostrea rhizophorae* to represent the entire resident molluscan assemblage. Living topographically higher than the tidally submerged Mangrove Oysters, on the overhanging tree branches and leaves, is the other conspicuous component of this

Figure 2.5 Close-up of an aggregation of the Mangrove Oyster, *Crassostrea rhizophorae* (Guilding, 1828), attached to a Red Mangrove prop root in Lake Worth. Specimens from this locality are particularly beautiful, having well-developed fluted spines on their shells. Red Mangrove forests are components of the Vegetated Sediment Shore Macrohabitat and support the *Crassostrea rhizophorae* Assemblage. (Photograph by Eddie Matchett.)

estuarine molluscan assemblage, the Mangrove Periwinkle, *Littoraria angulifera* (Figure 2.6 and 2.7A, B). Although also occurring on rocky outcrops, piers, and pilings, the Mangrove Periwinkle prefers living on Red Mangrove branches, where it scrapes algal films from the surface of the bark.

Larger clumps of Mangrove Oysters, particularly those at the bases of the prop roots, create the habitat for several conspicuous macromollusks. Within the labyrinthine cavities between the interconnected oyster valves, several species of mytilids occur, with *Brachidontes exustus* and *Ischadium recurvus* the most frequently encountered (Figures 2.7D, F). These mussels attach themselves to the oyster valves with strong byssal threads and often occur together with large aggregations of the encrusting anomiid bivalve *Anomia simplex* (Figure 2.7E). The sessile mytilid bivalves are sympatric with the filter-feeding calyptraeid gastropods *Ianacus atrasolea* and *Crepidula convexa* (Figures 2.7G and 2.8C–E), and these all form a distinctive suspension-feeding deme or subcomponent of the *Crassostrea rhizophorae* Assemblage. Within the cavities of the larger clumps of oysters, two carnivorous gastropods are frequently encountered: the muricid *Stramonita floridana* (illustrated and discussed in Chapter 3), which feeds on the oysters by drilling holes through their shells, and the Measled Cowrie, *Macrocypraea zebra* (Figure 2.8A, B), which is a generalist feeder and omnivore, preferring sponges, hydroids, carrion, or algae. Small specimens of the Measled Cowrie often occur within dead valves of the oyster *Crassostrea virginica* (Figure 2.8F), which frequently form large biohermal oyster banks around the base of the Red Mangroves or within channels between individual mangrove forests.

Another conspicuous faunal component of the *Crassostrea rhizophorae* Assemblage is the neritid gastropod *Nerita (Theliostyla) fulgurans*, which occurs throughout the prop root oyster clumps and adjacent hard surfaces (Figure 2.8G, H). The species is particularly common in brackish water areas of estuaries, where it feeds on surficial algal films growing on oyster shells and exposed limestone rocks, often in direct sunlight. In contrast, the smaller mangrove neritid *Nerita (Theliostyla) lindae* is far more cryptic and photophobic, completely shunning direct sunlight and often burying itself in mud beneath clumps of *Crassostrea virginica*. As in the case of its larger relative, the smaller *Nerita (Theliostyla) lindae* (Figure 2.8I, J) feeds on algal films and is most common along the Palm Beach coast, where it is sympatric with *Nerita (Theliostyla) fulgurans* in the Lake Worth Lagoon. *Nerita (Theliostyla) lindae* has also been reported from Biscayne Bay, Barnes Sound, and Black Water Sound and from the coral rock shoreline of Scout Key near Bahia Honda Key.

Biodiversity of the Crassostrea rhizophorae Assemblage

Like the Black Mangrove forests, the Red Mangrove community is subject to extremes in temperature, salinity, and desiccation, often during a single day's time span. Because of these harsh factors, the molluscan fauna of the *Crassostrea rhizophorae* Assemblage is relatively impoverished, incorporating only seven species of macrogastropods and eight of macrobivalves. These are listed here by feeding type and ecological niche.

1. **HERBIVORES (including algivores)**
   Gastropoda
      Neritidae
         Nerita (Theliostyla) fulgurans Gmelin, 1791 (Figure 2.8G, H)
         Nerita (Theliostyla) lindae Petuch, 1988 (Figure 2.8I, J)
      Littorinidae
         Littoraria angulifera (Lamarck, 1822) (Figure 2.7A, B)

Figure 2.6 Close-up of living Mangrove Periwinkles, *Littoraria angulifera* (Lamarck, 1822), crawling on a Red Mangrove branch on Missouri Key, Middle Florida Keys. These conspicuous arboreal gastropods are characteristic of the *Crassostrea rhizophorae* Assemblage. (Photograph by E. J. Petuch.)

Figure 2.7 Mollusks of the *Crassostrea rhizophorae* Assemblage. A, B = *Littoraria angulifera* (Lamarck, 1822), length 27 mm. C = *Crassostrea rhizophorae* (Guilding, 1828), length 43 mm. D = *Brachidontes exustus* (Linnaeus, 1758), length 24 mm. E = *Anomia simplex* d'Orbigny, 1853, width 28 mm. F = *Ischadium recurvum* (Rafinesque, 1820), length 29 mm. G = Close-up of the interior of a *Crassostrea rhizophorae* valve, showing the calyptraeid gastropods *Ianacus atrasolea* (large white specimens) and *Crepidula convexa* (small brown specimens) in life position.

Figure 2.8 Mollusks of the *Crassostrea rhizophorae* Assemblage. A, B = *Macrocypraea zebra* (Linnaeus, 1758), length 61 mm. C = *Crepidula convexa* Say, 1822, length 6 mm. D, E = *Ianacus atrasolea* (Collin, 2002), length 28 mm. F = *Crassostrea virginica* (Gmelin, 1791), clump length 106 mm. G, H = *Nerita (Theliostyla) fulgurans* Gmelin, 1791, width 28 mm. I, J = *Nerita (Theliostyla) lindae* Petuch, 1988, width 15 mm.

2. CARNIVORES (including scavengers and obligate omnivores)
Gastropoda
Cypraeidae
Macrocypraea zebra (Linnaeus, 1758) (Figure 2.8A, B)
Muricidae-Rapaninae
Stramonita floridana (Conrad, 1837)
3. SUSPENSION/FILTER FEEDERS
Gastropoda
Calyptraeidae
Crepidula convexa Say, 1822 (Figure 2.8C)
Ianacus atrasolea (Collin, 2002) (Figure 2.8D, E)
Bivalvia
Mytilidae
Brachidontes domingensis (Lamarck, 1819)
Brachidontes exustus (Linnaeus, 1758) (Figure 2.7D)
Ischadium recurvum (Rafinesque, 1820) (Figure 2.7F)
Ostreidae
Crassostrea rhizophorae (Guilding, 1828) (Figure 2.7C)
Crassostrea virginica (Gmelin, 1791) (Figure 2.8F)
Ostreola equestris Say, 1834
Anomiidae
Anomia simplex d'Orbigny, 1853 (Figure 2.7E)
Dreissenidae
Mytilopsis sallei (Recluz, 1849)

See Abbott (1974) and Warmke and Abbott (1962) for photographs of species not illustrated in this book. A review of the niche preferences of the members of the *Crassostrea rhizophorae* Assemblage shows that filter feeders dominate the Red Mangrove forests.

Molluscan ecology of the supratidal hypersaline and brackish pools

In the Florida Keys, Florida Bay, and Ten Thousand Islands, the supratidal area landward of the Black Mangrove forests normally contains large expanses of both hypersaline evaporite (Figure 2.9) and brackish water pools. These muddy, organic-rich transitory wetlands receive inputs of salt water only during the highest spring tides or during violent storms, and the water chemistry of this impounded environment fluctuates wildly throughout the year. During the rainy summer months, diluted brackish water conditions predominate, while during the dry winter months, highly concentrated saline conditions prevail. In some cases, the shallow muddy pools actually dry out completely, creating thick crusts of evaporite salts and forming sabkha-like conditions. Because of these extreme environmental parameters, with wildly shifting salinities and temperatures, only a few mollusks have evolved the physiological mechanisms to flourish in such a dystrophic system.

Two of the dominant vegetation types found near the hypersaline pools are the Saltworts (*Salicornia*), which comprise a species complex of at least three taxa (*Salicornia bigelovii, Salicornia maritima,* and *Salicornia herbacea*), and the Sea Purslane (*Sesuvium portulacastrum*). These succulent plants provide both the substrate and the habitat for a small assemblage of semiterrestrial gastropods, principal among these being the truncatellids

Figure 2.9 View of a large hypersaline pool on Ohio Key, Middle Florida Keys. Salt ponds such as this often house immense aggregations of small bivalves such as the corbiculid *Polymesoda maritima* and the corbulid *Caryocorbula contracta*. These provide a major food resource for seabirds and migrating water fowl. Supratidal salt ponds and saltwort thickets are components of the Vegetated Sediment Shore Macrohabitat and support the *Cerithideopsis costatus* Assemblage. (Photograph by Eddie Matchett.)

Truncatella pulchella and *Truncatella caribaeensis* and the potamidids *Cerithideopsis costatus* (Figure 2.10A) and *Cerithideopsis scalariformis* (Figure 2.10B, C). The cryptic truncatellid species live buried in moist leaf litter, where they feed on dead plant matter. The potamidids, on the other hand, live exposed on the *Salicornia* and *Sesuvium* stems and feed on encrusting algal films. The ellobiid gastropods *Melampus coffeus* and *Melampus bidentatus* occasionally may occur along with the *Cerithideopsis* species, particularly in supratidal hypersaline pools that are very near, or are interspersed between Black Mangrove and Green Buttonwood (*Podocarpus erectus*) forests. Only three species of bivalves are known to occur in the Florida Keys hypersaline and brackish pools: the corbiculid *Polymesoda maritima*, the dreissenid mussel *Mytilopsis leucophaeta* (Figure 2.10L), and the small corbulid *Caryocorbula contracta*. These bivalves, which have adapted to the harsh environmental conditions, often occur in vast aggregations and serve as a major food source for water fowl. Since the ecologically tightly restricted potamidid *Cerithideopsis costatus* is one of the most abundant and conspicuous mollusks of the hypersaline pool environs, we have selected this taxon to represent the entire resident molluscan assemblage.

The higher, dry land areas immediately adjacent to the hypersaline pools, referred to here as the **Terrestrial Interface**, are typically covered by dense growths of Salt Grass (*Distichus spicata*), Seashore Paspalum (*Paspalum vaginatum*), Seashore Dropseed (*Sporobolus virginicus*), and Cordgrass (*Spartina patens*). These beach grasses, along with the creeping tendrils of the Railroad Vine (*Ipomoea pes-caprae*; Figure 2.11) and the Beach Morning Glory (*Ipomoea imperati*), shelter a complex of salt-tolerant land snails in the genus *Cerion*. At least four subspecies of the endemic Florida Keys species *Cerion incanum* are known to exist: *Cerion incanum incanum* (Figure 2.10D, E), which is found from Key Largo south to Key West; *Cerion incanum vaccinum* (Figure 2.10J, K), which is found from Vaca Key south to No Name Key; *Cerion incanum saccharimeta* (Figure 2.10F, G), which is found on Sugar Loaf Key north to No Name Key; and *Cerion incanum fasciatum* (Figure 2.10H, I), which is endemic to Key Biscayne. These tropical Terrestrial Interface *Cerion* species belong to a group that is restricted to coastal southeastern Florida, the Bahamas, the West Indian Arc, and the Dutch West Indies (ABC Islands), and the Keys species complex is near the extreme northwestern edge of the range of the genus. Although classified as land snails, most *Cerion* species live on beach grasses near the shoreline, where they are continuously under the influence of salt spray.

Biodiversity of the Cerithideopsis costatus Assemblage

Only seven species of macromollusks remain when the cerionid land snails are excluded from the complete list of supratidal hypersaline and brackish pool molluscan inhabitants. This extreme impoverishment is a direct reflection of the harsh and continuously mutable environmental conditions of the Saltwort and Sea Purslane thickets and muddy saline marshlands. The members of the *Cerithideopsis costatus* Assemblage are listed here by feeding type and ecological niche.

1. **HERBIVORES (including algivores and leaf litter feeders)**
   Gastropoda
   Potamididae
   Cerithideopsis costatus (da Costa, 1778) (Figure 2.10A)
   Cerithideopsis scalariformis (Say, 1825) (Figure 2.10B, C)
   Truncatellidae
   Truncatella caribaeensis Reeve, 1842
   Truncatella pulchella Pfeiffer, 1839

Figure 2.10 Mollusks of the *Cerithideopsis costatus* Assemblage and Terrestrial Interface. A = *Cerithideopsis costatus* (da Costa, 1778), length 11 mm. B, C = *Cerithideopsis scalariformis* (Say, 1825), length 19 mm. D, E = *Cerion incanum incanum* (Binney, 1851), length 22 mm. (Missouri Key) F, G = *Cerion incanum saccharimeta* Pilsbry and Vanatta, 1899, length 29 mm (Sugar Loaf Key). H, I = *Cerion incanum fasciatum* (Binney, 1859), length 23 mm (Key Biscayne). J, K = *Cerion incanum vaccinum* Pilsbry, 1902, length 27 mm (No Name Key). L = *Mytilopsis leucophaeta* (Conrad, 1831), length 26 mm.

Figure 2.11 View of living specimens of the endemic Florida Keys Peanut Snail, *Cerion incanum incanum* (Binney, 1851), crawling on a Railroad Vine (*Ipomoea pes-caprae*) on Missouri Key. These salt-tolerant land snails live in close proximity to the beach and are characteristic of the Terrestrial Interface along the entire Florida Keys island chain. (Photograph by Eddie Matchett.)

2. CARNIVORES

No molluscan carnivores; replaced by the Raccoons *Procyon lotor auspicatus*, *Procyon lotor incautus*, *Procyon lotor inesperatus*, and *Procyon lotor marinus* and water fowl; on the larger mangrove islands of the Key West National Wildlife Refuge, the Mangrove Terrapin *Malaclemys terrapin rhizophorarum* was observed to feed on *Cerithideopsis scalariformis*.

3. SUSPENSION/FILTER FEEDERS

Bivalvia
Corbiculidae
Polymesoda maritima (d'Orbigny, 1842) (*P. floridana* is a synonym) (Figure 2.12)
Dreissenidae
Mytilopsis leucophaeta (Conrad, 1831) (Figure 2.10L)
Corbulidae
Caryocorbula contracta (Say, 1822)

TERRESTRIAL INTERFACE

4. HERBIVORES (including algivores)

Gastropoda
Cerionidae
Cerion incanum incanum (Binney, 1851) (Figure 2.10D, E)
Cerion incanum fasciatum (Binney, 1859) (Figure 2.10H, I)
Cerion incanum saccharimeta Pilsbry and Vanatta, 1899 (Figure 2.10F, G)
Cerion incanum vaccinum Pilsbry, 1902 (Figure 2.10J, K)

See Abbott (1974) and Warmke and Abbott (1962) for photographs of species not illustrated in this book. A review of the niche preferences of the members of the *Cerithideopsis costatus* Assemblage shows that molluscan herbivores dominate the supratidal hypersaline and brackish pools.

Another ecosystem that is unique to the Lower Florida Keys is one composed of **Dwarf Mangrove Meadows** that grow primarily along the northern sides of the Oölite Keys, from Big Pine Key west to at least Key West (Figure 2.13). The hard oölitic limestone shelves that skirt these islands are covered with only a few centimeters of fine carbonate mud, and the Red and Black Mangroves that grow there are stunted from the lack of soil and are, essentially, naturally-occurring bonsai trees. Some of these mangroves may be over fifty years in age but are only a meter or so in height, and these create a unique Lilliputian forest that extends for kilometers. The intertidal meadows and dwarf forests support an impoverished malacofauna composed of only a few hardy species, ones that can tolerate extremes in temperature and salinity. Typically, aggregations of the purple-pink corbiculid bivalve *Polymesoda martima* literally carpet the algae-covered carbonate mud, often forming a solid pavement of imbedded individuals (Figure 2.12). These small bivalves, in turn, are the principal prey items for a distinctive dwarf form of the Keys endemic melongenid, *Melongena (Rexmela) bicolor*. The branches, prop roots, and pneumatophores of the bonsai mangroves also support a rich fauna of ellobiid and potamidid gastropods, including *Melampus morrisoni*, *M. monilis*, *M. coffeus*, *Cerithideopsis scalariformis*, and *C. costatus*.

Figure 2.12 Close-up of an open carbonate sediment area between bonsai mangroves in a Dwarf Mangrove Meadow off Middle Torch Key, Lower Florida Keys. Here, the small purple-pink corbiculid bivalve *Polymesoda maritima* can be seen to form a virtual pavement, with many individuals being embedded or partially-embedded in the algae-covered carbonate sediment. A dwarf individual of the endemic Florida Keys Crown Conch, *Melongena (Rexmela) bicolor*, can be seen crawling at the lower center. (Photograph by Howard Peters).

Figure 2.13 View of a Dwarf Mangrove Meadow, composed of naturally-occurring bonsai Red Mangroves, growing along the northern coast of the Saddlebunch Keys, Lower Florida Keys. The open areas between the mangroves house large aggregations of the corbiculid bivalve *Polymesoda maritima* and dwarf forms of the Keys endemic melongenid gastropod *Melongena* (*Rexmela*) *bicolor*. (Photograph by E.J. Petuch).

chapter three

Molluscan faunas of the unvegetated rocky shore macrohabitat (nearshore marine regime)

Introduction

The seaward coasts of all the keys, from Soldier Key on Biscayne Bay south to Garden Key in the Dry Tortugas, are edged with narrow platforms of exposed, unvegetated carbonate rocks (Figure 3.1). From Soldier Key to Bahia Honda Key, these rock platforms are composed of outcrops of the Key Largo Formation and are made up entirely of weathered and recrystallized heads of massive fossil corals. These exposed fossil reefs are the remnants of the extensive late Pleistocene coralline systems that underlie the Upper and Middle Florida Keys (discussed in Chapter 1). From Scout Key (formerly West Summerland Key) and Big Pine Key westward to Key West and Boca Grande Key, these exposed rock platforms are composed of outcrops of the Miami Formation and are made up of oölitic limestone, also of late Pleistocene age (see Chapter 1). These eroded limestone outcrops, with their microkarstic surfaces, were aptly described by Rachel Carson when she pointed out, "On some of the Keys the rock is smoothly weathered, with flattened surfaces and rounded contours, but on many others the erosive action of the sea has produced a rough and deeply pitted surface, reflecting the solvent action of centuries of waves and driven salt spray. It is almost like a stormy sea frozen into solidity" (1955, p. 206).

These stark, eroded limestone outcrops, which often extend from the subtidal (sublittoral) zone all the way up to the supratidal (supralittoral) and terrestrial zones, support three separate zonated molluscan assemblages, with each restricted to only one narrow band. The vertically separated faunules include the *Cenchritis muricatus* Assemblage (highest supratidal zone), the *Nerita versicolor* Assemblage (intertidal zone), and the *Cerithium lutosum* Assemblage (shallow subtidal zone and tidal pools). Despite the harsh ecological conditions and physiological constraints placed on the biota of these tidally influenced limestone shores, a rich and distinctive molluscan fauna has managed to flourish. Although covered with films of green and blue-green filamentous algae, these rocky shelves characteristically lack large complex algae or other more advanced plants.

Molluscan ecology of the supratidal rocky outcrops

The area of the supratidal zone represents the topographically highest margin of the marine environment. Here, the rocky environment is splashed with salt water only during the highest spring tides and during strong storms, and the ecological parameters are essentially terrestrial. In the Florida Keys and along the Palm Beach coast, this interface

Figure 3.1 View of the ocean side of Missouri Key, Middle Florida Keys, showing a typical Florida Keys rocky tide pool shoreline. This type of environment is a component of the Unvegetated Rocky Shore Macrohabitat and supports the *Cenchritis muricatus* Assemblage, the *Nerita versicolor* Assemblage, and the *Cerithium lutosum* Assemblage. (Photograph by Eddie Matchett.)

zone between the sea and the land characteristically supports five species of extremely hardy, highly sculptured littorinid gastropods (known as "prickly winkles"). The ornate beads, spines, and knobs seen on their shells aid in keeping the animals cooler, both by reflecting the intense midday sunlight and by dissipating the baking tropical heat of this harsh exposed rocky habitat. Living in the highest area above the water line, and flourishing within this extreme environment, is the largest of the supratidal littorinids, *Cenchritis muricatus* (Figure 3.2). This heavily beaded species, with its reflective pale shell color, is abundant on the driest rocks and is frequently found crawling on Sea Purslane (*Sesuvium portulacastrum*) plants along with the salt-tolerant land snails of the genus *Cerion* (see Chapter 2). As the most conspicuous and characteristic gastropod of the rocky supratidal zone, we have selected this distinctive littorinid to represent the entire molluscan faunule (as the *Cenchritis muricatus* Assemblage).

Living slightly lower on the exposed supratidal rocks than *Cenchritis muricatus* (Figure 3.3A, B) are four other species of smaller prickly winkles: three species in the genus *Echinolittorina* and one in the genus *Tectininus*. In the Florida Keys, *Echinolittorina tuberculata* (Figures 3.3E and 3.4) is the most commonly encountered member of its genus, occurring in abundance in eroded cavities and pits on the exposed rock surfaces. Along the rocky coquina limestone coastlines of Palm Beach County (made up of outcrops of the late Pleistocene Anastasia Formation), the larger *Echinolittorina antonii* (Figure 3.3F, G) is the most abundant species of the genus and frequently occurs along with the small, extremely spiny *Echinolittorina dilatata* (Figure 3.3H, I). The proportionally wider, flatter-shelled *Tectininus nodulosus* (Figure 3.3C, D), with its two rows of large spines, is the most distinctive-appearing prickly winkle and ranges along the entire Palm Beach coastline and Florida Keys island chain. All five of these highly specialized littorinids feed on algal films that grow within small pits and crevices on the eroded limestone outcrops.

Biodiversity of the Cenchritis muricatus *Assemblage*

The extreme environmental conditions of the highest rocky supralittoral zone prevent the establishment of a diverse molluscan fauna. Because of this, only five algivorous gastropods occur here, forming a very impoverished molluscan faunule. The members of the *Cenchritis muricatus* Assemblage consist of only one gastropod family; these taxa are listed next:

1. **HERBIVORES (including algivores)**
   Gastropoda
   Littorinidae
       Cenchritis muricatus (Linnaeus, 1758) (Figure 3.3A, B)
       Echinolittorina antonii (Philippi, 1846) (Figure 3.3F, G)
       Echinolittorina dilatata (d'Orbigny, 1841) (Figure 3.3H, I)
       Echinolittorina tuberculata (Menke, 1828) (Figures 3.3E and 3.4)
       Tectininus nodulosus (Pfeiffer, 1839) (Figure 3.3C, D)

A review of the niche preferences of the members of the *Cenchritis muricatus* Assemblage shows that molluscan herbivores dominate the supratidal rocky outcrops. As is typical of stressed ecosystems, the low species diversity seen here is offset by a high frequency of individuals. Most often, all five of these species occur by the thousands on any small area of rocky coastline.

Figure 3.2 Close-up of living specimens of the littorinid gastropod *Cenchritis muricatus* (Linnaeus, 1758) in the highest supratidal area of the rocky shoreline on Missouri Key, Middle Florida Keys. These conspicuous semiterrestrial gastropods are characteristic of the *Cenchritis muricatus* Assemblage of the Unvegetated Rocky Shore Macrohabitat. (Photograph by Eddie Matchett.)

Figure 3.3 Littorinid gastropods of the *Cenchritis muricatus* Assemblage. A = *Cenchritis muricatus* (Linnaeus, 1758), length 14 mm. B = *Cenchritis muricatus* (Linnaeus, 1758), highly colored variant, length 15 mm. C = *Tectininus nodulosus* (Pfeiffer, 1839), length 12 mm. D = *Tectininus nodulosus* (Pfeiffer, 1839), length 10 mm. E = *Echinolittorina tuberculata* (Menke, 1828), length 10 mm. F = *Echinolittorina antonii* (Philippi, 1846), length 12 mm. G = *Echinolittorina antonii* (Philippi, 1846), form *major* Usticke, 1969, length 15 mm. H = *Echinolittorina dilatata* (d'Orbigny, 1841), length 10 mm. I = *Echinolittorina dilatata* (d'Orbigny, 1841), form *thiarella* Anton, 1838, length 11 mm.

Figure 3.4 Close-up of living specimens of the littorinid gastropod *Echinolittorina tuberculata* (Menke, 1828) in the high supratidal area of the rocky shoreline on Missouri Key, Middle Florida Keys. These small, ornately sculptured gastropods are typical components of the *Cenchritis muricatus* Assemblage of the Unvegetated Rocky Shore Macrohabitat. (Photograph by Eddie Matchett.)

Molluscan ecology of the littoral rocky outcrops

The intertidal environment of the Littoral Zone offers a more stable set of ecological parameters than does the topographically higher supratidal area. Here, the resident marine organisms are exposed, alternately, to dry subaerial conditions twice a day during low tides and to total submergence in seawater twice a day during high tides. This rhythmic alternation of water levels has allowed for more predictable and less-extreme conditions to prevail and has allowed for the evolution of a much richer malacofauna. The most conspicuous group of gastropods found in this intermediate zone is the Neritidae (Nerite Shells), which often cover the rock surfaces in large aggregations (Figure 3.5). In the Florida Keys and along the Palm Beach coast, the most abundant nerite species is *Nerita (Ritena) versicolor*; this large and colorful intertidal gastropod is here chosen to be the namesake of the entire molluscan faunule (the *Nerita versicolor* Assemblage). Three of the neritids found in the *Nerita versicolor* Assemblage—*Nerita versicolor* itself (Figure 3.6A, B), *Nerita (Theliostyla) tessellata* (Figure 3.6C, D), and *Nerita (Linnerita) peloronta* (Figure 3.6E, F)—range throughout the Caribbean region, and their presence in Palm Beach and the Florida Keys marks their northernmost biogeographical extent along the North American coastline.

Although having a much smaller combined biomass, the intertidal periwinkles make up the most species-rich group of gastropods in the *Nerita versicolor* Assemblage, with eight species in four different genera. Unlike the ornately sculptured shells of the supratidal littorinids, with their spines and knobs, the shells of the intertidal littorinids are relatively smooth, only faintly sculptured with fine spiral grooves. The predominant higher taxon of littorinids in the intertidal zone, the genus *Amerilittorina*, is composed of four sympatric species: *A. angustior* (Figure 3.6G, H), *A. jamaicensis* (Figure 3.6I, J), *A. ziczac* (Figure 3.6K), and *A. placida* (Figure 3.6L). These four species make up the bulk of the littorinid biomass and occur together with less-common, tiny littorinids such as *Fossarilittorina meleagris* (Figure 3.7A) and *Melarhaphe mespillum* (Figure 3.7E) and other rock-dwelling gastropods, such as the small planaxid *Angiola lineata* (Figure 3.7B) and the air-breathing false limpets *Siphonaria (Patellopsis) pectinata* (Figure 3.7C, D) and *Siphonaria (Patellopsis) alternata* (Figure 3.7F, G). All of these members of the *Nerita versicolor* Assemblage are grazers on algal films, and their sympatry and great frequency of individuals can be explained by a nonlimiting supply of algae food resources, preventing competition between species. This is a perfect example of the effects and interactions outlined in the Volterra–Gauss principle, by which competition can be overridden by an abundance of food resources.

The large herbivorous component of the *Nerita versicolor* Assemblage is, itself, the principal prey for a small contingent of carnivorous-molluscivorous gastropods in the family Muricidae. These highly specialized muricids, *Stramonita rustica* (Figure 3.7H), *Stramonita floridana* (Figure 3.7I, J), and *Plicopurpura patula* (Figure 3.7K, L), feed almost exclusively on the littorinids and planaxids, drilling holes through their thin shells and rasping out the tissues with the file-like radula. The more eurytopic *Stramonita* species also occur in shallow subtidal depths, where they feed primarily on small oysters and other sessile bivalves. In the Florida Keys, the *Stramonita* species live highest up on the rock faces, slightly above the mean tide line, and feed on small periwinkles such as *Amerilittorina angustior*, *Amerilittorina jamaicensis*, and *Amerlittorina ziczac*. The larger, flattened *Plicopurpura patula* is adapted for more turbulent water conditions and is found at, or slightly below, the mean tide line, where it is often exposed to wave action. There, it feeds on larger littorinids such as *Littoraria nebulosa* and its color form *tesselata* and the large planaxid *Planaxis (Supplanaxis) nucleus*.

Figure 3.5 Close-up of living specimens of the neritid gastropod *Nerita* (*Ritena*) *versicolor* Gmelin, 1791, in the rocky littoral zone on Missouri Key, Middle Florida Keys. These conspicuous gastropods are characteristic of the *Nerita versicolor* Assemblage of the Unvegetated Rocky Shore Macrohabitat. (Photography by Eddie Matchett.)

Figure 3.6 Gastropods of the *Nerita versicolor* Assemblage. A, B = *Nerita (Ritena) versicolor* Gmelin, 1791, length 20 mm. C, D = *Nerita (Theliostyla) tessellata* Gmelin, 1791, length 18 mm. E, F = *Nerita (Linnerita) peloronta* Linnaeus, 1758, length 24 mm. G, H = *Amerilittorina angustior* (Mörch, 1876), length 8 mm. I, J = *Amerilittorina jamaicensis* (C.B. Adams, 1850), length 9 mm. K = *Amerilittorina ziczac* (Gmelin, 1791), length 12 mm. L = *Amerilittorina placida* (Reid, 2009), length 3 mm.

Figure 3.7 Gastropods of the *Nerita versicolor* Assemblage. A = *Fossarilittorina meleagris* (Potiez and Michaud, 1838), length 5.2 mm. B = *Angiola lineata* (da Costa, 1778), length 6 mm. C, D = *Siphonaria (Patellopsis) pectinata* (Linnaeus, 1758), length 18 mm. E = *Melarhaphe mespillum* (Megerle von Mühlfeld, 1824), length 4 mm. F, G = *Siphonaria (Patellopsis) alternata* Say, 1826, length 9 mm. H = *Stramonita rustica* (Lamarck, 1822), length 27 mm. I, J = *Stramonita floridana* (Conrad, 1837), length 33 mm. K, L = *Plicopurpura patula* (Linnaeus, 1758), length 35 mm.

Biodiversity of the Nerita versicolor *Assemblage*

Although containing a richer malacofauna than that of the supratidal areas, the intertidal zone still houses a relatively impoverished molluscan assemblage, with only around 20 gastropods and no bivalves. These are listed here by feeding type and ecological niche.

    1. HERBIVORES (including algivores)

      Gastropoda

      Neritidae

          Nerita (Theliostyla) tessellata Gmelin, 1791 (Figure 3.6C, D)

          Nerita (Ritena) versicolor Gmelin, 1791 (Figure 3.6A, B)

          Nerita (Linnerita) peloronta Linnaeus, 1758 (Figure 3.6E, F)

          Puperita pupa (Linnaeus, 1767) (Middle and Lower Florida Keys only)

      Littorinidae

          Amerilittorina angustior (Mörch, 1876) (Figure 3.6G, H)

          Amerilittorina jamaicensis (C.B. Adams, 1850) (Figure 3.6I, J)

          Amerilittorina placida (Reid, 2009) (Figure 3.6L)

          Amerilittorina ziczac (Gmelin, 1791) (Figure 3.6K)

          Fossarilittorina meleagris (Potiez and Michaud, 1838) (Figure 3.7A)

          Littoraria nebulosa (Lamarck, 1822) (and color form *tessellata* Philippi, 1847)

          Melarhaphe mespillum (Mühlfeld, 1824) (Figure 3.7E)

      Planaxidae

          Angiola lineata (da Costa, 1778) (Figure 3.7B)

          Planaxis (Supplanaxis) nucleus (Bruguiere, 1789)

      Siphonariidae

          Siphonaria (Patellopsis) alternata (Linnaeus, 1758) (Figure 3.7F, G)

          Siphonaria (Patellopsis) pectinata Say, 1826 (Figure 3.7C, D)

          Williamia krebsii (Mörch, 1877)

    2. CARNIVORES (including drilling molluscivores)

      Muricidae-Rapaninae

          Plicopurpura patula (Linnaeus, 1758) (Figure 3.7K, L)

          Stramonita floridana (Conrad, 1837) (Figure 3.7I, J)

          Stramonita rustica (Lamarck, 1822) (Figure 3.7H)

A review of the niche preferences of the members of the *Nerita versicolor* Assemblage shows that molluscan herbivores dominate the rocky intertidal zone.

Molluscan ecology of the sublittoral rocky outcrops

The rocky intertidal and subtidal pools of the Florida Keys island chain are unique in the eastern United States in that they constitute the only hard substrate tide pool environment south of New England. Unlike the cold-water, impoverished malacofauna of the rocky New England coast (part of the Acadian Molluscan Province), the rocky coast of the Florida Keys houses a rich tropical fauna that contains a number of endemic taxa. The most prominent of these endemic Keys species is the dwarf crown conch, *Melongena (Rexmela) bicolor* (Figure 3.8I, J) (and its color form *estephomenos*; Figures 3.8A and 3.9). Most frequently encountered in rocky pools at low tide, this small melongenid is a voracious predator, feeding on the small cerithiid and batillariid gastropods that carpet the rock

Figure 3.8 Mollusks of the *Cerithium lutosum* Assemblage. A = *Melongena (Rexmela) bicolor* (Say, 1827), form *estephomenos* (Melvill, 1881), length 33 mm. B = *Collisella leucopleura* (Gmelin, 1791), view of shell interior, length 11 mm. C, D = *Trachypollia nodulosa* (C.B. Adams, 1845), length 17 mm. E = *Barbatia cancellaria* (Lamarck, 1811), length 29 mm. F = *Arcopsis adamsi* (Dall, 1886), length 12 mm. G, H = *Prunum apicinum virgineum* (Jousseaume, 1875), length 11 mm. I, J = *Melongena (Rexmela) bicolor* (Say, 1827), typical color form, length 37 mm.

surfaces, often by the thousands. The endemic Florida Keys dwarf crown conch ranges from Biscayne Bay south to the Dry Tortugas and is occasionally found in exposed pools in Florida Bay. Farther north of these areas, *Melongena* (*Rexmela*) *bicolor* is replaced by other congeneric taxa, such as *Melongena* (*Rexmela*) *corona corona* (Ten Thousand Islands northward to Cedar Key, western Florida; see Chapter 7) and *Melongena* (*Rexmela*) *corona winnerae* (Palm Beach to the St. Lucie River estuary; see Chapter 6). The extremely abundant prey item of the dwarf crown conch, the small rock-dwelling cerithiid *Cerithium lutosum*, is here chosen as the namesake of the entire rocky subtidal molluscan faunule (the *Cerithium lutosum* Assemblage).

The undersides of large rocks exposed in the tide pools act as refuges for a large fauna of fissurellid and lottiid limpets, one of the richest known from the western Atlantic. These algivorous species have 11 species in six different genera: *Diodora*, *Fissurella*, and *Lucapina* (Fissurellidae) and *Collisella*, *Lottia*, and *Patelloida* (Lottiidae). Although the limpets remain hidden at low tide, they are active at high tide, often grazing on algal films several meters from their home rocks. Some of the more conspicuous and commonly encountered limpets in the rocky tide pool areas include the fissurellids (keyhole limpets) *Diodora cayenensis* (Figure 3.10A), *Diodora listeri* (Figure 3.10B), *Diodora minuta* (Figure 3.10C), *Lucapina sowerbii* (Figure 3.10D), and *Lucapina suffusa* (Figure 3.10E) and the lottiids *Patelloida pustulata* (Figure 3.10K) and *Collisella leucopleura* (Figure 3.8B). Occurring under the rocks with the limpets is an interesting fauna of sessile ark shell bivalves, including the arcids *Cucullearca candida* (Figure 3.10J) and *Barbatia cancellaria* (Figure 3.8E) and the noetiid *Arcopsis adamsi* (Figure 3.8F). These small ark shells, which attach themselves to the underside of rocks with their tough byssal threads, are, along with the limpets, the principal prey items of the small, drilling carnivorous tide pool murex, *Trachypollia nodulosa* (Figure 3.8C, D). Specimens of the larger drilling molluscivorous muricids *Mancinella deltoidea* and *Strumonita floridana* also occur in these tide pool environments and compete with the small *Trachypollia* for sessile bivalve prey. Other small tide pool carnivores include the dove shells (Columbellidae) *Mitrella ocellata* (Figure 3.10I) and *Mitrella argus* (Figure 3.10H), which feed on sponges, hydroids, ectoprocts, and other small invertebrates.

One of the more incongruous members of the *Cerithium lutosum* Assemblage is the stocky, pure white marginellid gastropod *Prunum apicinum virgineum* (Figure 3.8G, H), a Florida Keys subspecies of the widespread Carolinian and Caribbean *Prunum apicinum* (North Carolina, the Gulf of Mexico, to the Virgin Islands). Most members of the family Marginellidae prefer soft sandy or muddy substrates, but this distinctive pure white *Prunum* is one of the few western Atlantic marginellids that lives in rocky areas. On Missouri, Ohio, and Rachel Carson Keys, near Bahia Honda Key in the Middle Keys, *Prunum apicinum virgineum* is abundant in sand-filled tide pools along the rocky shorelines. This white Keys subspecies also occurs in the adjacent Turtle Grass beds but seems to prefer the tide pool areas, where it is most frequently encountered. Within these rocky pools, this small marginellid forms large aggregations of hundreds of individuals, particularly when attracted to the presence of carrion, its favorite food source. Further studies, particularly DNA analyses, of the wide-ranging amphiprovincial *Prunum apicinum* may show that this common intertidal marginellid actually represents a complex of several closely related species, with *virgineum* being endemic to the Florida Keys.

Figure 3.9 Close-up of living specimens of the melongenid gastropod *Melongena (Rexmela) bicolor* (Say, 1827) form *estephomenos* (Melvill, 1881) in a tidal pool, at low tide, on Missouri Key, Middle Florida Keys. These endemic gastropods are typical components of the *Cerithium lutosum* Assemblage of the Unvegetated Rocky Shore Macrohabitat and are one of the main predators on cerithiid and batillariid gastropods. (Photograph by Eddie Matchett.)

Figure 3.10 Mollusks of the *Cerithium lutosum* Assemblage. A = *Diodora cayenensis* (Lamarck, 1822), length 19 mm. B = *Diodora listeri* (d'Orbigny, 1842), length 22 mm. C = *Diodora minuta* (Lamarck, 1822), length 11 mm. D = *Lucapina sowerbii* (Sowerby II, 1835), length 17 mm. E = *Lucapina suffusa* (Reeve, 1850), length 15 mm. F, G = *Cerithium lutosum* Menke, 1828, length 11 mm. H = *Mitrella argus* d'Orbigny, 1842, length 6 mm. I = *Mitrella ocellata* (Gmelin, 1791), length 10 mm. J = *Cucullearca candida* (Helbling, 1779), length 34 mm. K = *Patelloida pustulata* (Helbling, 1779), length 17 mm.

Biodiversity of the Cerithium lutosum *Assemblage*

Although not species rich, the rocky subtidal pools contain the most ecologically diverse faunule found in the entire Florida Keys rocky shoreline areas. The members of the *Cerithium lutosum* Assemblage are listed here by feeding type and ecological niche.

1. HERBIVORES (including algivores)
  Fissurellidae
    Diodora cayenensis (Lamarck, 1822) (Figure 3.10A)
    Diodora listeri (d'Orbigny, 1842) (Figure 3.10B)
    Diodora minuta (Lamarck, 1822) (Figure 3.10C)
    Fissurella barbadensis (Gmelin, 1791)
    Fissurella rosea (Gmelin, 1791)
    Lucapina sowerbii (Sowerby II, 1835) (Figure 3.10D)
    Lucapina suffusa (Reeve, 1850) (Figure 3.10E)
  Lottiidae
    Collisella leucopleura (Gmelin, 1791) (Figure 3.8B)
    Lottia antillarum (Sowerby I, 1831) (rare in the Florida Keys; see Abbott, 1974)
    Patelloida pustulata (Helbling, 1779) (Figure 3.10K)
  Cerithiidae
    Cerithium lutosum Menke, 1828 (Figure 3.10F, G)
  Batillariidae
    Batillaria minima (Gmelin, 1791) (eurytopic; most abundant on mud flats)
2. SUSPENSION/FILTER FEEDERS
  Gastropoda
  Vermetidae
    Dendropoma irregulare (d'Orbigny, 1841)
    Petaloconchus erectus (Dall, 1888)
  Bivalvia
  Arcidae
    Acar domingensis (Lamarck, 1819)
    Barbatia cancellaria (Lamarck, 1811) (Figure 3.8E)
    Cucullearca candida (Helbling, 1779) (Figure 3.10J)
  Noetiidae
    Arcopsis adamsi (Dall, 1886) (Figure 3.8F)
3. CARNIVORES (including scavengers)
  Muricidae-Rapaninae, Ergalitaxinae
    Mancinella deltoidea (Lamarck, 1822)
    Stramonita floridana (Conrad, 1837) (eurytopic; also in Littoral Zone)
    Trachypollia nodulosa (C.B. Adams, 1845) (Figure 3.8C, D)
  Melongenidae
    Melongena (*Rexmela*) *bicolor* (Say, 1827) (Figure 3.8I, J)
    Melongena (*Rexmela*) *bicolor* (Say, 1827) color form *estephomenos* (Melville, 1881)
      (Figure 3.8A)

Columbellidae
> *Mitrella argus* d'Orbigny, 1842 (Figure 3.10H)
> *Mitrella ocellata* (Gmelin, 1791) (Figure 3.10I)

Marginellidae
> *Prunum apicinum virgineum* (Jousseaume, 1875) (Figures 3.8G, H)

A review of the niche preferences of the members of the *Cerithium lutosum* Assemblage shows that there is a wide diversity of feeding types present, with 12 herbivores, 7 carnivores, and 5 filter feeders. The combined malacofauna of the rocky supralittoral, littoral, and shallow sublittoral zones (the *Cenchritis muricatus*, *Nerita versicolor*, and *Cerithium lutosum* Assemblages) of the Florida Keys is now known to total 49 species of gastropods and 3 species of bivalves.

Close-up of living *Cerithium lutosum* Menke, 1828 on a block of algae-covered fossil coral at Missouri Key, Middle Florida Keys, Florida. (Photograph by E.J. Petuch)

chapter four

Molluscan faunas of the vegetated softbottom macrohabitat (nearshore marine regime)

Introduction

The most frequently encountered biotopes in the shallow-water areas of the South Florida Bight, Florida Keys, and Florida Bay are the sea grass beds. These marine angiosperms cover around 75% of the open, shallow, soft-sediment seafloors and create a set of habitats that harbor some of the richest molluscan faunules found in southern Florida. Composed of three genera of sea grasses—*Thalassia* (Turtle Grass), *Halodule* (Shoal Grass), and *Syringodium* (Manatee Grass)—the grass beds often completely carpet the seafloor, forming densely intertwined mats of rhizomes. These root mats provide shelter for a species-rich assemblage dominated by shallowly burrowing bivalves. The three sea grass genera segregate themselves by bathymetry, with *Halodule* preferring intertidal mud flats and *Thalassia* and *Syringodium* preferring deeper-water areas (1- to 10-m depths, depending on water clarity). Because of these bathymetric differences, two different molluscan assemblages occur in the Vegetated Softbottom Macrohabitat: the *Bulla occidentalis* Assemblage (intertidal *Halodule* beds) and the *Modulus calusa* Assemblage (deeper-water *Thalassia* beds). As between all the biotopes covered in this book, these two main habitat types interfinger and often blur together in broad ecotonal transition zones.

Molluscan ecology of the Shoal Grass beds

Closest to the shore, and immediately adjacent to the Red Mangrove forests, beds of Shoal Grass (generally *Halodule wrightii*) often cover large areas of the intertidal mudflats (Figure 4.1). Although exposed at low tide and often subject to extremes in environmental conditions, these sea grass beds house a rich molluscan fauna, including a number of endemic species. These muddy intertidal flats support a rich and interesting fauna of carnivorous gastropods, comprising 10 species of thin-shelled bubble shells in the families Bullidae, Haminoeidae, and Aplustridae. Some of these include the large, dark brown, and mottled *Bulla occidentalis* (Figure 4.2J, K); the green-colored *Haminoea antillarum* (Figure 4.2H, I); the brown-colored *Haminoea taylorae* (endemic to Palm Beach and the Florida Keys; Figure 4.2F, G); the exquisite, net-patterned *Micromelo undatus* (Figures 4.2A, B and 4.3); and the large striped *Hydatina vesicaria* (Figures 4.2L and 4.4). These occur along with the fasciolariid *Cinctura hunteria* (Figure 4.2C, D), the melongenid *Melongena (Rexmela) bicolor* (Figure 4.2E), the cystiscid *Granulina hadria*, and the small raphitomids *Pyrgocythara plicosa* and *Pyrgocythara filosa*. Because *Bulla occidentalis* is the most frequently encountered and most conspicuous mollusk in the Shoal Grass beds, we have chosen it to represent the entire faunule (the *Bulla occidentalis* Assemblage).

Figure 4.1 View of a typical Shoal Grass (*Halodule wrightii*) bed at low tide in Lake Worth, Palm Beach County. Note the dead specimen of the bubble shell *Bulla occidentalis* A. Adams, 1850, an abundant gastropod that is characteristic of the Shoal Grass beds of southern Florida. Likewise, the small, black eurytopic batillariid gastropod, *Batillaria minima* (Gmelin, 1791), is abundant here but also occurs on open intertidal mudflats and rocky tide pools. This type of environment supports the *Bulla occidentalis* Assemblage of the Vegetated Softbottom Macrohabitat. (Photograph by Eddie Matchett.)

Figure 4.2 Gastropods of the *Bulla occidentalis* Assemblage. A, B = *Micromelo undatus* (Bruguiere, 1792), length 10 mm. C, D = *Cinctura hunteria* (Perry, 1811), length 48 mm. E = *Melongena (Rexmela) bicolor* (Say, 1827), dark color form, length 33 mm. F, G = *Haminoea taylorae* Petuch, 1987, length 10 mm. H, I = *Haminoea antillarum* (d'Orbigny, 1841), length 8 mm. J, K = *Bulla occidentalis* A. Adams, length 27 mm. L = *Hydatina vesicaria* (Lightfoot, 1786), length 38.4 mm.

Figure 4.3 View of the living animal of the aplustrid bubble shell, *Micromelo undatus* (Bruguiere, 1792), crawling in sand and shell rubble within the Lake Worth Lagoon, Palm Beach County, Florida. This delicate, beautifully patterned gastropod is a typical component of the *Bulla occidentalis* Assemblage of the Vegetated Softbottom Macrohabitat. (Photograph by Robert Myers.)

Figure 4.4 View of the living animal of the large aplustrid bubble shell, *Hydatina vesicaria* (Lightfoot, 1786), crawling in sand within the Lake Worth Lagoon, Palm Beach County, Florida. This distinctive banded gastropod is a typical component of the *Bulla occidentalis* Assemblage of the Vegetated Softbottom Macrohabitat. (Photograph by Robert Myers.)

The bubble shell species feed on a variety of small invertebrate prey, including the hydroids and ectoprocts that attach to the grass blades and small bivalves, gastropods, and crustaceans. Thin-shelled small bivalves, such as the mytilid *Arcuatula papyria* (Figure 4.5A) and the tellinid *Macoma cerina* (Figure 4.5E) are the frequent prey of the large *Bulla occidentalis*, while larger resident bivalves such as the lucinid *Phacoides pectinata* (Figure 4.5C), the deep-burrowing venerid *Petricolaria pholadiformis* (Figure 4.5J) and pholadid *Barnea truncata* (Figure 4.5K), the semelid *Semele proficua* (Figure 4.5F), the psammobiid *Heterodonax bimaculatus* (Figure 4.5D), the carditid *Carditamera floridana* (Figure 4.5H), and the mactrid *Mactrotoma fragilis* (Figure 4.5I) are the principal prey items of the molluscivorous *Cinctura* and *Melongena*. The raphitomid gastropods, with their harpoon-like radular teeth, are specialized feeders, preying totally on the small polychaete worms that are shallowly buried among the Shoal Grass root system. In the Lake Worth Lagoon of Palm Beach County, small specimens of the conilithid gastropod *Jaspidiconus pfluegeri* (see Chapter 6) often occur on the intertidal mudflats of the *Bulla occidentalis* Assemblage. Like the related toxoglossate family Raphitomidae, these small cone shells hunt for polychaete worms and subdue them with their harpoon-like radular teeth.

Biodiversity of the Bulla occidentalis *Assemblage*

Although not as species rich or as ecologically diverse as the adjacent, deeper-water Turtle Grass beds, the Shoal Grass bed molluscan assemblage contains a number of distinctive elements. Primary among these is the unusually large fauna of bubble shells, comprising at least 11 species in four families and six genera. The acteocinid, bullid, haminoeid, and aplustrid bubble shells, together with at least 12 macrobivalves, dominate the *Halodule*-associated molluscan faunule. The members of the bubble shell-dominated *Bulla occidentalis* Assemblage listed by feeding preference and ecological niche follow.

1. HERBIVORES (including algivores)
Cerithiidae
> *Bittiolum varium* (Pfeiffer, 1840)

Batillariidae
> *Batillaria minima* (Gmelin, 1791) (eurytopic; found in several intertidal assemblages)

2. SUSPENSION/FILTER FEEDERS
Bivalvia
Arcidae
> *Lunarca ovalis* (Bruguiere, 1789)

Mytilidae
> *Arcuatula papyria* (Conrad, 1846) (Figure 4.5A)

Lucinidae
> *Divaricella dentata* (Wood, 1815)
> *Parvilucina costata* (d'Orbigny, 1842)
> *Phacoides pectinata* (Gmelin, 1791) (Figure 4.5C)

Carditidae
> *Carditamera floridana* Conrad, 1838 (Figure 4.5H)

Veneridae
> *Chione elevata* Say, 1822 (eurytopic on all softbottom environments; see Chapter 6)
> *Cyclinella tenuis* (Recluz, 1852)
> *Macrocallista nimbosa* (Lightfoot, 1786) (Figure 4.5G)
> *Petricolaria pholadiformis* (Lamarck, 1818) (Figure 4.5J)

Figure 4.5 Bivalves of the *Bulla occidentalis* Assemblage. A = *Arcuatula papyria* (Conrad, 1846), length 13 mm. B = *Mulinia lateralis* (Say, 1822), length 10 mm. C = *Phacoides pectinata* (Gmelin, 1791), length 30 mm. D = *Heterodonax bimaculatus* (Linnaeus, 1758), length 15 mm. E = *Macoma cerina* (C.B. Adams, 1845), length 9 mm. F = *Semele proficua* (Pulteney, 1799), length 29 mm. G = *Macrocallista nimbosa* (Lightfoot, 1786), length 81 mm. H = *Carditamera floridana* Conrad, 1838, length 22 mm. I = *Mactrotoma fragilis* (Gmelin, 1791), length 39 mm. J = *Petricolaria pholadiformis* (Lamarck, 1818), length 35 mm. K = *Barnea truncata* (Say, 1822), length 33 mm.

Tellinidae
> *Angulus sybariticus* (Dall, 1881)
> *Macoma cerina* (C.B. Adams, 1845) (Figure 4.5E)
> *Macoma constricta* (Bruguiere, 1792)

Semelidae
> *Semele proficua* (Pulteney, 1799) (Figure 4.5F)

Psammobiidae
> *Heterodonax bimaculatus* (Linnaeus, 1758) (Figure 4.5D)

Mactridae
> *Mactrotoma fragilis* (Gmelin, 1791) (Figure 4.5I)
> *Mulinia lateralis* (Say, 1822) (Figure 4.5B)

Corbulidae
> *Caryocorbula caribaea* (d'Orbigny, 1853)

Pholadidae
> *Barnea truncata* (Say, 1822) (Figure 4.5K)
> *Cyrtopleura costata* (Linnaeus, 1758) (illustrated in Chapter 6)

3. CARNIVORES (several specialized types)
3a. MOLLUSCIVORES

Fasciolariidae
> *Cinctura hunteria* (Perry, 1811) (Figure 4.2C, D)

Melongenidae
> *Melongena (Rexmela) bicolor* (Say, 1827) (Figure 4.2E; mudflat variant)

3b. VERMIVORES

Raphitomidae
> *Pyrgocythara filosa* Rehder, 1939
> *Pyrgocythara plicosa* (C.B. Adams, 1850)

3c. GENERAL CARNIVORES

Nassariidae
> *Phrontis vibex* (Say, 1822)

Cystiscidae-Granulininae
> *Granulina hadria* (Dall, 1889)

Haminoeidae
> *Atys riiseanus* Mörch, 1875
> *Atys sandersoni* Dall, 1881
> *Haminoea antillarum* (d'Orbigny, 1841) (Figure 4.2H, I)
> *Haminoea elegans* (Gray, 1825)
> *Haminoea succinea* (Conrad, 1846)
> *Haminoea taylorae* Petuch, 1987 (Figure 4.2F, G)

Bullidae
> *Bulla occidentalis* A. Adams, 1850 (Figure 4.2J, K)
> *Bulla solida* Gmelin, 1791

Aplustridae
> *Hydatina vesicaria* (Lightfoot, 1786) (Figures 4.2L and 4.4)
> *Micromelo undatus* (Bruguiere, 1792) (Figures 4.2A, B and 4.3)

Acteocinidae
> *Utriculastra canaliculata* (Say, 1822)

A review of the niche preferences of the members of the *Bulla occidentalis* Assemblage shows that carnivores dominate the faunule, with several specialized vermivorous and molluscivorous predators. This high diversity of feeding types is offset by the much larger biomass of the combined filter-feeding bivalve fauna.

Molluscan ecology of the Turtle Grass beds

In water depths greater than 0.5 m, and including the infralittoral subzone (areas exposed to subaerial conditions only during the lowest spring tides), the larger *Thalassia testudinum* beds begin to dominate the seafloor (Figure 4.6). In many localities, these Turtle Grass meadows grade slowly into the intertidal Shoal Grass beds in a broad area of transition that allows both species to live sympatrically. In these types of transitional zones, members of both the *Bulla occidentalis* Assemblage and the resident Turtle Grass–associated *Modulus calusa* Assemblage occur together in a rich ecotonal community. Also occurring with the *Thalassia* meadows in slightly deeper water is the Manatee Grass, *Syringodium*, which often grows together with the Turtle Grass to form dense thickets. In some areas with just the right bathymetric conditions, all three sea grass genera (*Halodule*, *Thalassia*, and *Syringodium*) can occur together, forming a rich and varied substrate for an exceptionally large assemblage of mollusks. Because the small, disk-shaped modulid gastropod *Modulus calusa* (Figure 4.7A–F) is one of the most conspicuous and dominant mollusks of the Turtle Grass beds, it has here been chosen to represent the entire malacofauna (the *Modulus calusa* Assemblage).

With 109 species of resident macromollusks, both bivalves and gastropods, the Turtle Grass beds shelter the third-largest molluscan assemblage found in the Florida Keys. Unlike the Shoal Grass beds, the *Thalassia*-based malacofauna contains a large component of algivorous herbivores, primarily in the families Lotiidae, Trochidae, Phasianellidae, Turbinidae, Litiopidae, Neritidae, Cerithiidae, Modulidae, Xenophoridae (which also feeds on foraminifera), and Strombidae. These algivores coexist with a very large and rich filter-feeding bivalve fauna, comprising 17 families, 41 genera, and over 50 species, and this is the second-largest bivalve assemblage found anywhere in the Florida Keys and Florida Bay (the open sand bottom *Polinices lacteus* Assemblage contains the largest number of bivalves; see Chapter 6). The large carnivore component of the Turtle Grass bed malacofauna comprises 44 species in 30 genera and 15 families and is one of the most ecologically diverse, encompassing general carnivores, scavengers, vermivores, echinoderm feeders, hydroid feeders, spongivores, and molluscivores.

Ecologically, the Turtle Grass beds separate into three distinct biotopes: the Individual Grass Blade Biotope, the Main Vegetative Mass Biotope of the grass bed, and the thickly tangled Rhizome Root Mat Biotope buried in the soft sediments. All three of these biotopes contain their own distinctive faunules. The individual grass blades act as demes, or separate microcommunities, and each houses an entire diminutive malacofauna. The main grass bed shelters many large gastropods and other invertebrates, shading the animals beneath the forest of grass blades and allowing them to avoid the detection of predators, such as stingrays (primarily Dasyatidae) and large snappers (Lutjanidae) and other perciform fish. The dense root mats provide a firm substrate for shallowly burrowing bivalves, many of which have become highly specialized and are restricted to Turtle Grass beds. The grass beds also house several endemic Florida Keys gastropods, primary among these the calliostomatid *Calliostoma adelae* (Figure 4.8D), the turbinid *Lithopoma americana* (Figure 4.8A), the modulid *Modulus calusa* (Figure 4.7A–F), the conid *Gradiconus burryae* (Figures 4.9A–F and 4.10A, B), and the conilithid *Jaspidiconus pealii* (Figure 4.9G–L).

Figure 4.6 View of a Turtle Grass (*Thalassia testudinum*) bed off Middle Torch Key, Lower Florida Keys, showing a resident Pinfish (*Lagodon rhomboides*). This type of sea grass environment supports the *Modulus calusa* Assemblage of the Vegetated Softbottom Macrohabitat. (Photograph by Ron Bopp.)

Figure 4.7 Gastropods of the *Modulus calusa* Assemblage. A, B = *Modulus calusa* Petuch, 1988, typical color form, width 13 mm. C = *Modulus calusa* Petuch, 1988, tan color form, width 12 mm. D = *Modulus calusa* Petuch, 1988, speckled color form, width 14 mm. E, F = *Modulus calusa* Petuch, 1988, banded color form, width 12 mm. This species has often been incorrectly referred to by many workers as *Modulus modulus* (Linnaeus, 1758), which is actually a different-looking, heavily knobbed species that is restricted to the southern Caribbean. G = *Cerithium muscarum* Say, 1822, length 24 mm. H = *Cerithium eburneum* Bruguiere, 1792, length 22 mm. I = *Cerithium atratum* (Born, 1778), length 32 mm. J = *Crepidula ustulatulina* Collin, 2002, on *Modulus calusa* Petuch, 1988, width 14.4 mm. K, L = *Crepidula ustulatulina* Collin, 2002, length 8.3 mm.

Figure 4.8 Gastropods of the *Modulus calusa* Assemblage. A = *Lithopoma americana* (Gmelin, 1791), height 24 mm (endemic to the Florida Keys). B = *Astralium phoebia* Röding, 1798, width 22 mm. C = *Turbo (Marmarostoma) castanea* (Gmelin, 1791), height 20 mm. D = *Calliostoma adelae* Schwengel, 1951, height 17.4 mm. E = *Tegula (Agathistoma) fasciata* (Born, 1778), width 13.7 mm. F = *Patelloida pulcherrima* (Petit, 1856), length 15 mm. G, H = *Cymatium femorale* (Linnaeus, 1758), length 138 mm. I, J = *Cymatium (Ranularia) cynocephalum* (Lamarck, 1816), length 68.5 mm. K = *Lobatus raninus* (Gmelin, 1791), length 97.5 mm. L = *Turbo (Taenioturbo) canaliculatus* Hermann, 1781, height 66 mm.

Figure 4.9 Cone Shells of the *Modulus calusa* Assemblage. A = *Gradiconus burryae* (Clench, 1942), typical form, length 35.5 mm. B = *Gradiconus burryae* (Clench, 1942), solid brown color form, length 22.7 mm. C = *Gradiconus burryae* (Clench, 1942), broad-shouldered variant, length 43 mm. D = *Gradiconus burryae* (Clench, 1942), dark color form, length 36 mm. E, F = *Gradiconus burryae* (Clench, 1942), pale color form, length 37.8 mm. G, H = *Jaspidiconus pealii* (Green, 1830), typical form, length 13.2 mm. I = *Jaspidiconus pealii* (Green, 1830), banded color form, length 13.2 mm. J = *Jaspidiconus pealii* (Green, 1830), yellow color form, length 16 mm. K = *Jaspidiconus pealii* (Green, 1830), orange color form, length 17 mm. L = *Jaspidiconus pealii* (Green, 1830), blue-and-green color form, length 12 mm.

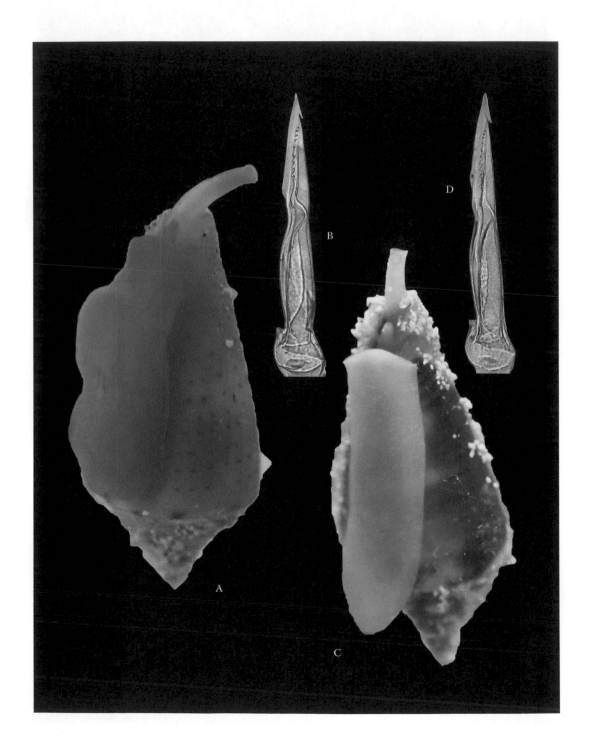

The individual grass blade demes house entire microcosms, with their broad, flat surfaces characteristically encrusted with coralline algae, pennate diatoms, and other microalgae; ectoproct bryozoans (typically the cheilostome *Schizoporella floridana*); sponges; colonial urochordate tunicates; and pinnatulid hydroids. These microcommunities support a large number of small, grass blade–dwelling gastropods, such as the carnivorous columbellid dove shells *Costoanachis avara* (Figure 4.11F), *Costoanachis floridana* (Figure 4.11D), *Columbella rusticoides* (Figure 4.11A), and *Zafrona taylorae* (Figure 4.11B), which feed on hydroids, tunicates, and small invertebrates; the highly specialized drilling molluscivore naticid *Haliotinella patinaria*; the tiny scavenger marginellids *Dentimargo aureocinta*, *Volvarina garycooverti*, and *Volvarina avena*; the colorful persiculine cystiscids *Persicula pulcherrima* and *Gibberula catenata*; the vermivorous cone shell *Jaspidiconus pealii*; and grazing algivores such as the limpet *Patelloida pulcherrima* (Figure 4.8F), the trochid *Tegula fasciata* (Figure 4.8E), the phasianellid *Eulithidium thalassicola*, the tiny green neritid *Smaragdia viridemaris*, the tiny litiopod *Alaba incerta*, and the modulid *Modulus calusa*. Frequently, the dorsum of the Calusa Button Shell, *Modulus calusa*, serves as the substrate for the attachment of the small filter-feeding calyptraeid *Crepidula ustulatulina* (Figure 4.7J–L). These two small gastropods form an odd symbiosis that is unique to the sea grass beds of Florida Bay and the Keys. A number of small but distinctive bivalves also attach to the individual grass blades and are important components of the deme community. Some of these include the pectinid scallops *Antillipecten antillarum* (Figure 4.12I, J) and *Lindapecten exasperatus* (Figure 4.12G, H) and the pteriid pearl oyster *Pinctada longisquamosa* (Figure 4.13B, C). These small sessile bivalves, which are attached to the grass blades by their byssal threads, serve as prey items for the voracious molluscivorous gastropods *Haliotinella patinaria* and the six species of columbellids.

The densely packed and tightly entangled Turtle Grass blades form huge undersea meadows that often extend, uninterrupted, for kilometers. Larger carnivorous gastropods, such as the molluscivorous fasciolariids *Triplofusus papillosus* (Figures 4.14H and 4.15) and *Fasciolaria tulipa* (Figure 4.14G); the echinoderm-feeding ranellids *Cymatium femorale* (Figure 4.8G, H) and *Cymatium (Ranularia) cynocephalum* (Figure 4.8I, J); the drilling molluscivores *Chicoreus dilectus* (Figure 4.14D–F) and *Phyllonotus pomum* (Figure 4.14A–C); the buccinid *Hesperisternia multangula* (Figure 4.14I); and the vermivorous conid *Gradiconus burryae*, all find shelter here and are protected from fish predation. Also sheltering among the grass

Figure 4.10 Living Animals of Florida Keys *Gradiconus* species. A = *Gradiconus burryae* (Clench, 1942), pale orange color form found in Turtle Grass beds off the eastern side of Missouri Key, Middle Florida Keys. Note the bright orange animal; shell length 23 mm. B = Radular tooth of *Gradiconus burryae* (Clench, 1942); taken from the 23-mm Missouri Key specimen. C = *Gradiconus mazzolii* (Petuch and Sargent, 2011), from the sponge bioherms off Middle Torch Key, Lower Florida Keys. Note the lemon yellow animal; shell length 21 mm (discussed and illustrated in Chapter 5). D = Radular tooth of *Gradiconus mazzolii* (Petuch and Sargent, 2011); taken from the 21-mm Middle Torch Key specimen. These two closely related species were considered to be conspecific by several cone workers, but the animals are very different in coloration and live in very different habitats. The harpoon-like radular teeth of both species also differ from each other: The tooth of *Gradiconus burryae* is proportionally thicker and has a wide, projecting lateral blade and a small terminal hook; the tooth of *Gradiconus mazzolii* is proportionally more slender, with a narrow, reduced lateral blade and a larger terminal hook. Along with the offshore, coral reef-associated *Gradiconus anabathrum tranthami* (Petuch, 1995) (Figure 6.10A–D, Chapter 6; which has a salmon-pink animal), these characteristic Florida Keys species constitute an endemic species radiation that is unique in the Americas. (Live animal photographs by William Bennight; photomicrographs of radular teeth by Dr. Manuel Tenorio.)

Figure 4.11 Gastropods of the *Modulus calusa* Assemblage. A = *Columbella rusticoides* Heilprin, 1886, length 14.7 mm. B = *Zafrona taylorae* Petuch, 1987, length 10 mm (endemic to the Florida Keys). C = *Costoanachis sertulariarum* (d'Orbigny, 1839), length 11 mm. D = *Costoanachis floridana* Rehder, 1939, length 13 mm. E = *Costoanachis sparsa* (Reeve, 1859), length 11 mm. F = *Costoanachis avara* (Say, 1822), length 10 mm. G, H = *Prunum carneum* (Storer, 1837), length 20 mm. I, J = *Prunum guttatum* (Dillwyn, 1817), dark color form, length 22 mm. K, L = *Prunum guttatum* (Dillwyn, 1817), pale color form, length 21 mm.

Figure 4.12 Pectinid bivalves of the *Modulus calusa* Assemblage. A, B = *Argopecten irradians taylorae* Petuch, 1987, length 34 mm. C, D = *Argopecten irradians taylorae* Petuch, 1987, orange color form, length 35 mm. E, F = *Argopecten nucleus* (Born, 1778), length 32 mm. G, H = *Lindapecten exasperatus* (Sowerby II, 1842), length 22 mm (= *acanthodes* Dall). I, J = *Antillipecten antillarum* (Recluz, 1853), length 19 mm.

Figure 4.13 Bivalves of the *Modulus calusa* Assemblage. A = *Arcopagia fausta* (Pulteney, 1799), length 51 mm. B = *Pinctada longisquamosa* (Dunker, 1852), length 35 mm. C = *Scissula similis* (Sowerby I, 1806), length 21 mm. D = *Tucetona pectinata* (Gmelin, 1791), length 21 mm. E = *Tellinella listeri* (Röding, 1798), length 47 mm. F = *Ctena orbiculata* (Montagu, 1808), length 11 mm. G = *Chione mazyckii* Dall, 1902, length 21 mm. H = *Pitar fulminatus* (Menke, 1828), length 42 mm. I = *Periglypta listeri* (Gray, 1838), length 54 mm.

Figure 4.14 Gastropods of the *Modulus calusa* Assemblage. A, B = *Phyllonotus pomum* (Gmelin, 1791), length 69.4 mm. C = *Phyllonotus pomum* (Gmelin, 1791), length 67 mm. D, E = *Chicoreus dilectus* (A. Adams, 1855), length 49 mm. F = *Chicoreus dilectus* (A. Adams, 1855), dark color form, length 55 mm. G = *Fasciolaria tulipa* (Linnaeus, 1758), length 77 mm. H = *Triplofusus papillosus* (Sowerby I, 1825), length 386 mm. I = *Hesperisternia multangula* (Philippi, 1848), length 22.4 mm. J = *Eupleura sulcidentata* Dall, 1890, length 16 mm. K = *Calotrophon ostrearum* (Conrad, 1846), length 18 mm.

Figure 4.15 Close-up of a living Horse Conch, *Triplofusus papillosus* (Sowerby I, 1825), in a Turtle Grass bed off Missouri Key, Middle Florida Keys. Note the characteristic bright orange-red animal. This species is the Florida State Shell and was formerly known as *Pleuroploca gigantea* Kiener, 1840, a name that is now considered to be a synonym. This gigantic mollusk, the largest gastropod in the Atlantic Ocean, is a common component of the *Modulus calusa* Assemblage of the Vegetated Softbottom Macrohabitat. (Photograph by Ron Bopp.)

blades and feeding on epiphytic algae is a large fauna of herbivorous gastropods, including the large turbinids *Turbo* (*Taenioturbo*) *canaliculatus* (Figure 4.8L) and *Turbo* (*Marmarostoma*) *castanea* (Figure 4.8C); the cerithiids *Cerithium atratum* (Figure 4.7I) and *Cerithium muscarum* (Figure 4.7G); and the strombid *Lobatus raninus* (Figure 4.8K). Smaller scavengers and carrion feeders, such as the marginellids *Prunum carneum* (Figure 4.11G, H), *Prunum guttatum* (Figure 4.11I–L), and *Prunum apicinum virgineum* and the nassariid *Phrontis vibex*, also occur within these dense sea grass meadows. The bulk of the gastropod species making up the *Modulus calusa* Assemblage, particularly the larger ones, are found in the main sections of the grass beds.

The open *Thalassia* meadows also serve as the substrate for immense aggregations of the pectinid scallop *Argopecten irradians taylorae* (Figure 4.12A–D), which continuously swim and migrate from one *Thalassia* thicket to another. This large pectinid, Taylor's Bay Scallop, has often been confused (i.e., Mikkelsen and Bieler, 2008) with the Eastern Bay Scallop, *Argopecten irradians concentricus*, a different subspecies that ranges from Cape Hatteras, North Carolina, to Fort Pierce, Florida. No bay scallops are found anywhere along the Palm Beach and Broward County coastlines of southern Florida, so there is a distinct biogeographical separation and barrier between the two subspecies (which differ in rib count, color, shell and rib shape, and DNA; see Marelli et al., 1997). *Argopecten irradians taylorae* first appears in Biscayne Bay, Dade County, and extends throughout the Florida Keys Turtle Grass beds and then northward up the western coast of Florida to the Mississippi coast. The brackish water Mississippi River Delta acts as another biogeographical barrier and has no resident bay scallops. Another subspecies, *Argopecten irradians amplicostatus*, appears along the Texas coast and extends to the Yucatan Peninsula of Mexico. North of Cape Hatteras, the cold-water subspecies *Argopecten irradians irradians* ranges to Cape Cod, the biogeographical terminus of the species complex (Petuch, 2013: 17). In some areas of Florida Bay, such as in the Rabbit Key Basin, *Argopecten irradians taylorae* is so abundant that, in the 1980s, it formed the basis of a commercial scalloping industry (officially stopped by Everglades National Park in 1985). Another related scallop, *Argopecten nucleus* (Figure 4.12E, F), also occurs along with *Argopecten irradians taylorae* in the Florida Bay and Florida Keys Turtle Grass beds but is a much rarer species, only sporadically encountered.

The Turtle Grass root mats serve as the preferred substrate for several large and characteristic sea grass bivalves, including the lucinids *Codakia orbicularis* (Figure 4.16A), *Lucina pensylvanica* (Figure 4.16B), and *Anodontia alba* (Figure 4.16C); the cardiids *Dallocardia muricata* (Figure 4.16D) and *Laevicardium serratum* (Figure 4.16E); the venerids *Dosinia discus* (Figure 4.16G), *Dosinia elegans* (Figure 4.16H), and *Periglypta listeri* (Figure 4.13I); the pinnids *Atrina rigida* (Figure 4.16J) and *Pinna carnea* (Figure 4.16I); the psammobiid *Sanguinolaria sanguinolenta*; and the tellinids *Arcopagia fausta* (Figure 4.13A), *Scissula similis* (Figure 4.13C), and *Tellinella listeri* (Figure 4.13E). Several highly specialized bivalves, such as the lyonsiids *Entodesma beana* and *Lyonsia floridana*, build nests at the bases of the Turtle Grass rhizomes and remain virtually undetectable. Smaller drilling carnivores, such as the muricids *Vokesimurex rubidus*, *Eupleura sulcidentata* (Figure 4.14J), and *Calotrophon ostrearum* (Figure 4.14K), all burrow among the Turtle Grass rhizomes, searching for small bivalve prey such as lyonsiids, mytilids, and small semelids.

Biodiversity of the Modulus calusa *Assemblage*

The rich malacofauna of the Turtle Grass beds of the Florida Keys also extends into deeper-water (2- to 5-m depths) areas of Florida Bay, where it inhabits the extensive meadows of Manatee Grass (*Syringodium filiforme*). These sea grass shoals, with individual grass blades

Figure 4.16 Bivalves of the *Modulus calusa* Assemblage. A = *Codakia orbicularis* (Linnaeus, 1758), length 63 mm. B = *Lucina pensylvanica* (Linnaeus, 1758), length 37 mm. C = *Anodontia alba* Link, 1807, length 49 mm. D = *Dallocardia muricata* (Linnaeus, 1758), length 31 mm. E = *Laevicardium serratum* (Linnaeus, 1758), length 30 mm. F = *Laevicardium mortoni* (Conrad, 1830), length 14 mm. G = *Dosinia discus* (Reeve, 1850), length 52 mm. H = *Dosinia elegans* (Conrad, 1846), length 48 mm. I = *Pinna carnea* Gmelin, 1791, length 98 mm. J = *Atrina rigida* (Lightfoot, 1786), length 142 mm.

over 1 m tall, are especially well developed in the areas of the Oxfoot, Schooner, Sprigger, and Tripod Banks and within the intervening "Sluiceway" channel (Thomas Frankovich, 2013 personal communication by email). The members of the *Thalassia* and *Syringodium*-dwelling *Modulus calusa* Assemblage are listed here by feeding type and ecological preference.

1. **HERBIVORES (including grazers on epiphytic algae)**
   Gastropoda
   Lotiidae
   > *Patelloida pulcherrima* (Petit, 1856) (Figure 4.8F)

   Trochidae
   > *Tegula (Agathistoma) fasciata* (Born, 1778) (Figure 4.8E)

   Turbinidae
   > *Astralium phoebiu* Röding, 1798 (Figure 4.8B)
   > *Lithopoma americana* (Gmelin, 1791) (Figure 4.8A)
   > *Turbo (Marmarostoma) castanea* (Gmelin, 1791) (Figure 4.8C)
   > *Turbo (Taenioturbo) canaliculatus* Hermann, 1781 (Figure 4.8L)

   Neritidae
   > *Smaragdia viridemaris* Maury, 1917

   Phasianellidae
   > *Eulithidium affine* (C.B. Adams, 1850)
   > *Eulithidium thalassicola* (Robertson, 1958)

   Litiopidae
   > *Alaba incerta* (d'Orbigny, 1841)

   Cerithiidae
   > *Cerithium atratum* (Born, 1778) (Figure 4.7I)
   > *Cerithium eburneum* Bruguiere, 1792 (Figure 4.7H)
   > *Cerithium muscarum* Say, 1822 (Figure 4.7G)

   Modulidae
   > *Modulus calusa* Petuch, 1988 (Figure 4.7A–F) (often incorrectly referred to as *Modulus modulus*, which is a southern Caribbean species that is not found in Florida)

   Strombidae
   > *Lobatus raninus* (Gmelin, 1791) (Figure 4.8K)

   Xenophoridae
   > *Xenophora conchyliophora* (Born, 1780) (feeds on both algae and foraminifera)

2. **CARNIVORES**
2a. **GENERAL CARNIVORES/SCAVENGERS**
   Buccinidae
   > *Hesperisternia multangula* (Philippi, 1848) (Figure 4.14I)

   Columbellidae
   > *Astyris lunata* (Say, 1826)
   > *Astyris multilineata* (Dall, 1889)
   > *Columbella rusticoides* Heilprin, 1886 (Figure 4.11A)
   > *Costoanachis avara* (Say, 1822) (Figure 4.11F)
   > *Costoanachis floridana* Rehder, 1939 (Figure 4.11D)
   > *Costoanachis sertulariarum* (d'Orbigny, 1839) (Figure 4.11C)
   > *Costoanachis sparsa* (Reeve, 1859) (Figure 4.11E)
   > *Parvanachis obesa* (C.B. Adams, 1845)
   > *Parvanachis ostreicola* (Sowerby II, 1882)
   > *Zafrona taylorae* Petuch, 1987 (Figure 4.11B)

Nassariidae
  Phrontis vibex (Say, 1822)
Marginellidae
  Dentimargo aureocincta Stearns, 1872
  Dentimargo eburneola Conrad, 1834
  Dentimargo idiochila Schwengel, 1943
  Dentimargo reducta (Bavay, 1922)
  Prunum apicinum virgineum (Jousseaume, 1875)
  Prunum carneum (Storer, 1837) (Figure 4.11G, H)
  Prunum guttatum (Dillwyn, 1817) (Figure 4.11I–L)
  Prunum succinea (Conrad, 1846) (= *veliei* Pilsbry)
  Volvarina avena (Kiener, 1834)
  Volvarina garycooverti Espinosa and Ortea, 1998
Cystiscidae-Persiculinae
  Gibberula catenata (Montagu, 1803)
  Persicula pulcherrina (Gaskoin, 1849)
Haminoeidae
  Haminoea antillarum (d'Orbigny, 1841)
  Haminoea succinea (Conrad, 1846)
  Haminoea taylorae Petuch, 1987
Bullidae
  Bulla occidentalis A. Adams, 1850 (Figure 4.2J, K)
2b. MOLLUSCIVORES
Naticidae
  Haliotinella patinaria (Guppy, 1876)
Muricidae-Muricinae, Ocenebrinae
  Calotrophon ostrearum (Conrad, 1846) (Figure 4.14K)
  Chicoreus dilectus (A. Adams, 1855) (Figure 4.14D–F)
  Eupleura sulcidentata Dall, 1890 (Figure 4.14J)
  Phyllonotus pomum (Gmelin, 1791) (Figure 4.14A–C)
  Vokesimurex rubidus (F.C. Baker, 1897)
Fasciolariidae
  Cinctura hunteria (Perry, 1811)
  Fasciolaria tulipa (Linnaeus, 1758) (Figure 4.14G)
  Triplofusus papillosus (Sowerby I, 1825) (Figure 4.14H)
2c. VERMIVORES
Conidae-Puncticulinae
  Gradiconus burryae (Clench, 1942) (Figures 4.9A–F and 4.10A, B)
  Lindaconus atlanticus (Clench, 1942)
Conilithidae-Conilithinae
  Jaspidiconus pealii (Green, 1830) (Figure 4.9G–L)
Drilliidae
  Cerodrillia thea (Dall, 1889) (rare in the Florida Keys; common along western Florida)
Raphitomidae
  Daphnella lymnaeiformis (Kiener, 1840)
  Pyrgocythara filosa Rehder, 1939
  Pyrgocythara hemphilli Bartsch and Rehder, 1939
  Stellatoma stellata (Stearns, 1872)

2d. ECHINODERM FEEDERS
Ranellidae
Cymatium femorale (Linnaeus, 1758) (Figure 4.8G, H)
Cymatium (Ranularia) cynocephalum (Lamarck, 1816) (Figure 4.8I, J)

3. SUSPENSION/FILTER FEEDERS
Gastropoda
Calyptraeidae
Crepidula ustulatulina Collin, 2002 (Figure 4.7J–L)
Bivalvia
Glycymeridae
Tucetona pectinata (Gmelin, 1791) (Figure 4.13D)
Mytilidae
Musculus laterulis (Say, 1822)
Pteriidae
Pinctada longisquamosa (Dunker, 1852) (Figure 4.13B)
Pinnidae
Atrina rigida (Lightfoot, 1786) (Figure 4.16J)
Atrina seminuda (Lamarck, 1819)
Atrina serrata (Sowerby I, 1825)
Pinna carnea Gmelin, 1791 (Figure 4.16I)
Pectinidae
Antillipecten antillarum (Recluz, 1853) (Figure 4.12I, J)
Argopecten irradians taylorae Petuch, 1987 (Figure 4.12A–D)
Argopecten nucleus (Born, 1778) (Figure 4.12E, F)
Lindapecten exasperatus (Sowerby II, 1842) (Figure 4.12G, H)
Crassitellidae
Crassinella lunulata (Conrad, 1834)
Carditidae
Carditamera floridana Conrad, 1838
Pteromeris perplana (Conrad, 1841)
Lyonsiidae
Entodesma beana (d'Orbigny, 1853)
Lyonsia floridana Conrad, 1849
Lucinidae
Anodontia alba Link, 1807 (Figure 4.16C)
Callucina keenae Chavan, 1971
Codakia orbicularis (Linnaeus, 1758) (Figure 4.16A)
Ctena orbiculata (Montagu, 1808) (Figure 4.13F)
Lucina pensylvanica (Linnaeus, 1758) (Figure 4.16B)
Lucinisca nassula (Conrad, 1846)
Parvilucina crenella (Dall, 1901)
Ungulinidae
Phlyctiderma semiaspera (Philippi, 1836)
Cardiidae
Dallocardium muricata (Linnaeus, 1758) (Figure 4.16D)
Laevicardium mortoni (Conrad, 1830) (Figure 4.16F)
Laevicardium serratum (Linnaeus, 1758) (Figure 4.16E)
Veneridae
Chione elevata Say, 1822

Chione mazyckii Dall, 1902 (Figure 4.13G)
Cyclinella tenuis (Recluz, 1852)
Dosinia discus (Reeve, 1850) (Figure 4.16G)
Dosinia elegans (Conrad, 1846) (Figure 4.16H)
Gouldia cerina (C.B. Adams, 1845)
Periglypta listeri (Gray, 1838) (Figure 4.13I)
Pitar fulminatus (Menke, 1828) (Figure 4.13H)
Pitar simpsoni (Dall, 1895)
Timoclea pygmaea (Lamarck, 1818)

Tellinidae
Angulus tampaensis (Conrad, 1866)
Arcopagia fausta (Pulteney, 1799) (Figure 4.13A)
Eurytellina alternata (Say, 1822)
Macoma brevifrons (Say, 1834)
Scissula candeana (d'Orbigny, 1853)
Scissula similis (Sowerby I, 1806) (Figure 4.13C)
Tellidora cristata (Recluz, 1842)
Tellinella listeri (Röding, 1798) (Figure 4.13E)

Psammobiidae
Sanguinolaria sanguinolenta (Gmelin, 1791)

Semelidae
Abra aequalis (Say, 1822)
Cumingia vanhyningi Rehder, 1939

Solecurtidae
Tagelus divisus (Spengler, 1794)
Tagelus plebeius (Lightfoot, 1786)

Mactridae
Spisula raveneli (Conrad, 1832)

A review of the niche preferences of the members of the *Modulus calusa* Assemblage shows that the faunule comprises a mixed ecosystem, with 16 herbivores, 44 carnivores, and 52 filter feeders. The filter feeders, mostly infaunal bivalves, dominate the assemblage and make up the largest biomass.

chapter five

Molluscan faunas of the unvegetated hardbottom macrohabitat (nearshore marine regime)

Introduction

In the shallow-water areas of the Florida Keys, particularly in the channels between the Lower Keys and along the smaller islands north of the Lower Keys, large expanses of eroded limestone are exposed within the sublittoral zone, usually at depths of 1–3 m. These heavily pitted carbonate rocks, with their microkarstic surfaces, are extensions of the islands themselves and are covered with a thin veneer of carbonate sand. This coarse surficial sediment layer, which constantly shifts with tidal movement, prevents the growth of sea grasses and macroalgae. As a result, only organisms that can attach themselves to the underlying solid rock surface can flourish in this type of environment. In other areas of the Keys, primarily along the seaward sides of the Upper and Middle Keys, these sublittoral rocky platforms are devoid of the thin carbonate sand layer. Here, rock-boring sea urchins have excavated large pits across the entire exposed rock surface, creating a unique hardbottom habitat of miniature caves and cavities.

The Unvegetated Hardbottom Macrohabitat, with its exposed rock pavements, houses three separate biotopes, each containing its own resident molluscan assemblage: the Sponge Bioherm Biotope ("sponge reefs"), with the *Cerodrillia clappi* Assemblage; the Gorgonian Forest Biotope, with the *Cyphoma rhomba* Assemblage; and the Rock-Boring Sea Urchin Microcave Biotope, with the *Bayericerithium litteratum* Assemblage. Several smaller, ancillary demes also occur along the sand-covered open rock seafloors, the most notable being the Yellow Mussel beds and their associated predatory muricid gastropod faunule. In many areas, particularly in the channels between the Lower Keys, all of these types of biotopes and their thin sand veneers (usually 1- to 10-cm thick) interfinger with Turtle Grass beds and form a reticulated, "patchwork quilt" arrangement of habitats.

Molluscan ecology of the sponge bioherms

The shallow subtidal (1- to 3-m depths) oölitic limestone platforms found along the northern sides of the Lower Florida Keys and within the intervening channels typically harbor large biohermal structures composed completely of various species of sponges. Many of these sponges, such as the Green Sponge (*Haliclona viridis*), the Loggerhead Sponge (*Spheciospongia vesparia*), the Vase Sponge (*Ircinia campana*; Figure 5.1), and the Cake Sponge (*Ircinia strobilina*), attach directly to the rock substrate; others, such as the Branching Candle Sponge (*Verongia longissima*) and the Sprawling Sponge (*Neopetrosia longleyi*), form ropy intertwined masses that either lie on top of the thin sand layer or coil around the larger attached species. On most of the sponge bioherms of the Lower Keys, at least 12

Figure 5.1 Close-up of a large Basket Sponge, *Ircinia campana*, growing in a sponge bioherm off Middle Torch Key, Lower Florida Keys. These sponge "reef" environments, which contain a large number of unusual endemic species, typically support the *Cerodrillia clappi* Assemblage of the Unvegetated Hardbottom Macrohabitat. (Photograph by Ron Bopp.)

to 14 different species of sponges occur together, producing a unique ecosystem unlike any other found in the southeastern United States. This sponge "reef" environment was recently found (Petuch and Sargent, 2011b) to contain a highly endemic malacofauna, also unlike any other in the Florida Keys or adjacent areas.

Although not particularly species rich, and containing only around 29 gastropods and 3 bivalves, the molluscan fauna of the sponge bioherms is highly localized and is restricted to the sand-covered oölitic limestone platforms. One of the most character-istic gastropods of the sponge bioherm areas is the Keys endemic drilliid gastropod, *Cerodrillia clappi* (Figure 5.2D), and we have chosen this ecologically restricted species to represent the entire molluscan faunule (the *Cerodrillia clappi* Assemblage; the similar *C. perryae* from western Florida and the Gulf of Mexico and the widespread *C. thea* are often misidentified as the Keys endemic *C. clappi*). This distinctive little endemic turroidean occurs together with a host of other sponge bioherm-associated endemic gastropod spe-cies, including the muricid *Vokesimurex rubidus* form *marcoensis* (Figure 5.3D, E), the nassariids *Uzita websteri* (Figure 5.3G, H) and *Uzita swearingeni* new species (described in the Systematic Appendix at the end of this book); the busyconid *Fulguropsis spiratum key-sensis* (Figure 5.4C); the slender, high-spired conid *Gradiconus mazzolii* (Figure 5.4F–K; Figure 4.10 in Chapter 4), and the bullid *Bulla frankovichi* (Figure 5.4D, E, L). These endemic taxa also occur together with a large complement of other toxoglossate gas-tropods, such as the crassispirid *Crassispira mesoleuca* (Figure 5.2B), the zonulispirids *Zonulispira crocata* and *Pilsbryspira leucocyma* (Figure 5.2F), the strictispirid *Strictispira redferni* (Figure 5.2C), the raphitomid *Pyrgocythara hemphilli* (Figure 5.2E), and the dril-liid *Neodrillia blacki* (Figure 5.2A). The species richness of this vermivorous turroi-dean fauna, along with the conoideans *Gradiconus mazzolii*, *Lindaconus atlanticus*, and *Jaspidiconus pealii*, indicates the presence of a diverse polychaete worm fauna with a large biomass.

The sponge masses themselves host a small but interesting molluscan fauna, com-prising one pectinid bivalve and several gastropods. Living near and within encrusting sponges such as *Haliclona rubens* and *Haliclona viridis*, and frequently encased entirely in a sponge coating, is the small spiny pectinid scallop, *Lindapecten muscosus* (Figure 5.4A, B). This characteristic sponge scallop comes in a variety of colors, ranging from red to brown and to bright yellow. Also buried in the sponge masses along with the sponge scallop is the small turritellid worm gastropod, *Vermicularia knorri* (Figure 5.3F). This sessile Keys sponge bioherm worm shell differs from typical specimens from elsewhere in Florida and the Caribbean by having a much smaller, more tightly coiled shell, and it may actually prove to represent a new endemic species or subspecies. Living on the surface of the large sponge species such as *Speciospongia vesparia* is a small fauna of algivorous gastropods. Primary among these are the turbinids *Lithopoma americana* and *Turbo* (*Marmarostoma*) *cas-tanea*, both of which also occur in adjacent Turtle Grass beds. These turbinids scrape algal films off the surface of the sponge pellis and frequently congregate in large numbers on Loggerhead Sponges.

Biodiversity of the Cerodrillia clappi *Assemblage*

The distinctive and characteristic sponge-associated malacofauna of the Lower Florida Keys, with its large number of endemic, ecologically restricted taxa, stands out as one of the most unusual in the tropical western Atlantic. The members of the oölitic limestone platform and sponge bioherm-dwelling *Cerodrillia clappi* Assemblage are listed here by feeding type and ecological niche preference.

Figure 5.2 Mollusks of the *Cerodrillia clappi* Assemblage and the Yellow Mussel beds. A = *Neodrillia blacki* Petuch, 2004, length 23.4 mm. This large drilliid possibly may be a shallow-water form or sub-species of the larger, deeper-water *Neodrillia moseri* (Dall, 1889). B = *Crassispira mesoleuca* Rehder, 1943, length 13.5 mm (endemic to the Florida Keys). C = *Strictispira redferni* Tippett, 2006, length 16.5 mm. D = *Cerodrillia clappi* Bartsch and Rehder, 1939, length 12 mm. E = *Pyrgocythara hemphilli* Bartsch and Rehder, 1943, length 8.2 mm. F = *Pilsbryspira leucocyma* (Dall, 1883), length 11 mm. G = *Favartia pacei* Petuch, 1988, length 15.6 mm. Animal and foot are pure white in color. H = *Favartia cellulosa* (Conrad, 1846), length 12 mm. Animal and foot are pale orange in color. I, J = *Murexiella caitlinae* Petuch, new species. Holotype, length 30 mm. Often referred to as *M. mcgintyi*, which is actually a late Pliocene–early Pleistocene fossil (see Systematic Appendix). K = *Murexiella glypta* (M. Smith, 1938), length 18 mm. L = *Brachidontes modiolus* (Linnaeus, 1758), length 29 mm. Varies in color from yellow to black.

Figure 5.3 Mollusks of the *Cerodrillia clappi* Assemblage. A = *Columbella mercatoria* (Linnaeus, 1758), length 13.6 mm. B = *Vexillum exiguum* (C.B. Adams, 1845), length 7 mm. (= *V. hanleyi* Dohrn, 1861) C = *Vexillum moniliferum* (C.B. Adams, 1850), length 9 mm. (= *V. albocinctum* C.B. Adams) D, E = *Vokesimurex rubidus* form *marcoensis* (Sowerby III, 1900), length 28.4 mm. (Note the bright purple aperture and high spire that are characteristic of this distinctive ecomorph). F = *Vermicularia knorri* (Deshayes, 1843), length 22 mm. G, H = *Uzita websteri* (Petuch and Sargent, 2011), holotype, length 8 mm. I = *Uzita consensa* (Ravenel, 1861), length 9 mm. J, K = *Uzita swearingeni* Petuch and Myers, new species, length 7 mm. (holotype; see Systematic Appendix) L = *Uzita swearingeni* Petuch and Myers, new species, length 8.5 mm.

1. **HERBIVORES (including algal film grazers)**
   Turbinidae
   > *Astralium phoebia* Röding, 1798
   > *Lithopoma americana* (Gmelin, 1791)
   > *Turbo (Marmarostoma) castanea* (Gmelin, 1791)
2. **CARNIVORES**
2a. **GENERAL CARNIVORES/SCAVENGERS**
   Nassariidae
   > *Uzita consensa* (Ravenel, 1861) (Figure 5.3I)
   > *Uzita swearingeni* Petuch and Myers, new species (Figure 5.3J–L) (see Systematic Appendix)
   > *Uzita websteri* (Petuch and Sargent, 2011) (Figure 5.3G, H)
   Columbellidae
   > *Columbella mercatoria* (Linnaeus, 1758) (Figure 5.3A)
   > *Columbella rusticoides* Heilprin, 1886
   Costellariidae
   > *Vexillum exiguum* (C.B. Adams, 1845) (Figure 5.3B) (= *hanleyi* Dohrn, 1861)
   > *Vexillum moniliferum* (C.B. Adams, 1850) (Figure 5.3C) (incorrectly referred to as *Vexillum albocinctum* C.B. Adams, which is a nomen dubium)
   Marginellidae
   > *Prunum apicinum virgineum* (Jousseaume, 1875)
   > *Prunum carneum* (Storer, 1837)
   > *Prunum guttatum* (Dillwyn, 1817)
   Bullidae
   > *Bulla frankovichi* Petuch and Sargent, 2011 (Figure 5.4D, E, L)
2b. **MOLLUSCIVORES (including drilling molluscivores)**
   Muricidae-Muricinae
   > *Chicoreus dilectus* (A. Adams, 1855)
   > *Vokesimurex rubidus* form *marcoensis* (Sowerby III, 1900) (Figure 5.3D, E)
   Fasciolariidae
   > *Fasciolaria tulipa* (Linnaeus, 1758)
   > *Dolicholatirus cayohuesonicus* (Sowerby II, 1878)

Figure 5.4 Mollusks of the *Cerodrillia clappi* Assemblage. A = *Lindapecten muscosus* (Wood, 1828), length 20 mm, yellow color form. B = *Lindapecten muscosus* (Wood, 1828), length 27 mm, normal color form. C = *Fulguropsis spiratum keysensis* Petuch, 2013, holotype, length 57.8 mm. This endemic Florida Keys subspecies of the Texan and Yucatanean *Fulguropsis spiratum spiratum* has the same aperture sculpture as the nominate subspecies but differs in having more numerous and much coarser spiral cords on the body whorl and spire. This is the most strongly sculptured and ornate member of the genus *Fulguropsis*. Compare with the smooth *Fulguropsis pyruloides* (Figure 7.14). D, E = *Bulla frankovichi* Petuch and Sargent, 2011, holotype, length 10.2 mm. (Note the slender shell shape, very narrow aperture, flat spire, and four dark brown color bands that distinguish this sponge bioherm-dwelling species from the much larger, mud-flat-dwelling *Bulla occidentalis*; see Figure 4.5A, B). F, G = *Gradiconus mazzolii* Petuch and Sargent, 2011. Holotype, length 16.7 mm. H = *Gradiconus mazzolii* Petuch and Sargent, 2011, length 20.8 mm. I = *Gradiconus mazzolii* Petuch and Sargent, 2011, dark color form, length 22 mm. J = *Gradiconus mazzolii* Petuch and Sargent, 2011, length 19.4 mm. K = *Gradiconus mazzolii* Petuch and Sargent, 2011, striped color form, 21 mm. For an illustration of the living animal of *Gradiconus mazzolii*, see Figure 4.10 in Chapter 4. L = *Bulla frankovichi* Petuch and Sargent, 2011, pale color variant, length 9 mm.

Busyconidae
  Fulguropsis spiratum keysensis Petuch, 2013 (Figure 5.4C)
2c. VERMIVORES
Conidae-Puncticulinae
  Gradiconus mazzolii Petuch and Sargent, 2011 (Figure 5.4F–K; Figure 4.10C, D in
    Chapter 4)
  Lindaconus atlanticus (Petuch, 1942)
Conilithidae-Conilithinae
  Jaspidiconus pealii (Green, 1830)
Drilliidae
  Cerodrillia clappi Bartsch and Rehder, 1939 (Figure 5.2D)
  Neodrillia blacki Petuch, 2004 (Figure 5.2A)
Strictispiridae
  Strictispira redferni Tippett, 2006 (Figure 5.2C)
Raphitomidae
  Pyrgocythara hemphilli Bartsch and Rehder, 1939 (Figure 5.2E)
Turridae
  Crassispira mesoleuca Rehder, 1943 (Figure 5.2B)
Zonulispiridae
  Pilsbryspira leucocyma (Dall, 1883) (Figure 5.2F)
  Zonulispira crocata (Reeve, 1845) (= *sanibelensis* Bartsch and Rehder, 1939)
3. SUSPENSION/FILTER FEEDERS
Gastropoda
Turritellidae
  Vermicularia knorri (Deshayes, 1843) (Figure 5.3F)
  Vermicularia spirata (Philippi, 1836)
Bivalvia
Pectinidae
  Lindapecten muscosus (Wood, 1828) (Figure 5.4A, B)
Glycymeridae
  Tucetona pectinata (Gmelin, 1791)
Tellinidae
  Scissula similis (Sowerby I, 1806)

A review of the niche preferences of the members of the *Cerodrillia clappi* Assemblage shows that gastropod carnivores dominate the ecosystem, with 26 species divided between general carnivores and scavengers and specialized feeders such as vermivores, and molluscivores.

Mollusks of the Yellow Mussel beds

Beds of the Yellow Mussel, *Brachidontes modiolus* (Figure 5.2L; varies in color from yellow to black), occur in scattered patches on open carbonate sand bottoms adjacent to the sponge bioherms and Turtle Grass meadows. These small mytilid bivalves form dense aggregations, with byssally anchored individuals literally touching one another and with just a small amount of fine sediment filling in between each shell. All of the individual mussels are oriented in the same direction, with the gaping posterior ends of their shells protruding slightly above the sediment layer in which they are buried. This almost solidly

packed amalgamation of individuals often covers many square meters of shallow seafloor and provides the habitat for several predatory muricid gastropods. Throughout the
Florida Keys, these buried mussel beds occur most frequently in channels and open sand
areas where there is a swift current during tidal changes. The best-developed mussel beds
have been found along the shorelines of southernmost Key Largo (Harry Harris Park),
Plantation Key, Vaca Key (Three Sisters Rocks), and Missouri Key.

The Yellow Mussels are subject to heavy predation by several small muricid gastropods, all of which drill holes through the shells of their bivalve prey with their specialized
radular teeth. Once the hole is drilled, the muricids insert their proboscises and feed on
the soft tissues of the mytilids, consuming the entire animal. Although some species can
occur in other environments throughout the Keys, three of the four muricids found on the
Brachidontes modiolus beds are most frequently encountered on the mytilid aggregations,
usually nestled down between the individual mussels. These include the following predatory species:

Gastropoda
Muricidae-Muricopsinae
  Favartia cellulosa (Conrad, 1846) (Figure 5.2H)
  Favartia pacei Petuch, 1988 (Figure 5.2G)
  Murexiella glypta (M. Smith, 1938) (Figure 5.2K)
  Murexiella caitlinae Petuch and Myers, new species (Figure 5.2I, J) (see Systematic
    Appendix)

Of special importance in this mytilid deme is the presence of the Keys endemic species
Murexiella caitlinae, which appears to be ecologically tightly bound to the Yellow Mussel
beds. This large and impressive muricid, which is found mostly in the *Brachidontes modiolus* beds, has been referred to by most previous authors as *Murexiella mcgintyi*. That species, however, is an ancestral fossil taxon that was originally described from the early
Pleistocene Caloosahatchee Formation of southern Florida (Gelasian Age). The living species and its fossil ancestor differ considerably in shell morphology, and these differences
are discussed in the Systematic Appendix at the end of this book. Another small sympatric
muricid, *Murexiella glypta*, was also named from earliest Pleistocene fossil specimens, and
the living representative does have a number of subtle morphological differences. Future
research, involving the comparison of the morphometrics of the living and fossil forms,
may show that the Recent Keys *glypta* actually represents another new, unnamed species.

Molluscan ecology of the gorgonian forests

On many of the exposed limestone platforms around the Keys, in depths of 1 to 3 m, an
unusually rich fauna of octocorallian cnidarians forms immense thickets, typically carpeting the seafloor (Figure 5.5). These shallow-water gorgonian forests are made up of as
many as 16 separate species, some of which include the Sea Whips *Pterogorgia anceps* and
Pterogorgia citrina, the Sea Plumes *Antillogorgia americana* and *Antillogorgia acerosa*, the Sea
Rod *Plexaura flexuosa*, the Corky Sea Fingers *Briareum asbestinum*, and Palmer's Eunicea
Eunicea palmeri, a species that is endemic to the Florida Keys area. The gorgonian thickets
of the Florida Keys house the largest shallow-water ovulid gastropod fauna found any-
where else in the Gulf of Mexico or Caribbean Sea. This remarkable fauna of ovulids ("Egg
Cowries") lives in close association with the gorgonians, feeding on the octocoral colonies

Figure 5.5 View of a gorgonian octocoral thicket off Dania Beach, Florida, showing the Sea Whips *Eunicea* and *Plexaurella*, the Sea Plume *Antillogorgia*, and the Sea Fan *Gorgonia ventalina*. These types of gorgonian aggregations house a large fauna of cnidarian-feeding ovulid gastropods, including endemic species such as *Cyphoma sedlaki* and *Cyphoma rhomba*. This type of environment supports the *Cyphoma rhomba* Assemblage of the Unvegetated Hardbottom Macrohabitat. (Photograph by Robert Myers.)

by cropping individual polyps. The ovulid gastropods move around the gorgonian colony, grazing on the polyps, while the octocoral colony regenerates the half-eaten polyps in a non-ending cycle. As is presently understood, eight different species of ovulids live on these gorgonian forests, including five species of the genus *Cyphoma* and one each of the genera *Pseudocyphoma*, *Cymbovula*, and *Simnialena*. We have chosen the endemic Florida Keys species *Cyphoma rhomba* (Figure 5.6G, H) to represent the entire molluscan faunule (as the *Cyphoma rhomba* Assemblage).

Living with *Cyphoma rhomba* on the gorgonians, often in abundance, are four other congeneric species (all referred to as "Flamingo Tongues"): the wide-ranging Caribbean *Cyphoma gibbosum* (Figure 5.6A–C) and *Cyphoma signatum* (Figure 5.6I), the widespread Carolinian Province *Cyphoma mcgintyi* (Figure 5.6D–F), and the tiny, rare Keys endemic *Cyphoma sedlaki* (Figure 5.6J). The color patterns of the exposed mantle tissue of the five Flamingo Tongues all differ greatly and can be used to tell the various species apart (see Figure 5.6 for illustrations of the preserved mantle color patterns). The multiple species of *Cyphoma* all live together with the rare Caribbean *Pseudocyphoma intermedium* (Figure 5.6K) and the small, thin, and delicate *Cymbovula acicularis* and *Simnialena uniplicata*. Attached to the different gorgonian species by hook-like shell prongs is the ostreid bivalve *Dendostrea frons* (Figure 5.6L), often referred to as the "Gorgonian Oyster" or "Coon Oyster." These bizarre, heavily rippled oysters conform to the shape and thickness of their resident gorgonian, often taking on an elongated body form.

Biodiversity of the Cyphoma rhomba *Assemblage*

All the members of the gorgonian-dwelling *Cyphoma rhomba* Assemblage live in close association with the octocoral colonies, either as predators on the living polyps or as ectocommensals that attach themselves to the gorgonian stems. These are listed here by feeding type and ecological niche preference.

1. **CARNIVORES (including cnidarian feeders)**
   Gastropoda
   Ovulidae
   > *Cymbovula acicularis* (Lamarck, 1810)
   > *Cyphoma gibbosum* (Linnaeus, 1758) (Figure 5.6A–C)
   > *Cyphoma mcgintyi* Pilsbry, 1939 (Figure 5.6D–F)
   > *Cyphoma rhomba* Cate, 1978 (Figure 5.6G, H)
   > *Cyphoma sedlaki* Cate, 1976 (Figure 5.6J)
   > *Cyphoma signatum* Pilsbry and McGinty, 1939 (Figure 5.6I)
   > *Pseudocyphoma intermedium* (Sowerby II, 1828) (Figure 5.6K)
   > *Simnialena uniplicata* (Sowerby II, 1849)
2. **SUSPENSION/FILTER FEEDERS**
   Bivalvia
   Pteriidae
   > *Pteria colymbus* (Röding, 1798) (attached to gorgonians)
   Ostreidae
   > *Dendostrea frons* (Linnaeus, 1758) (attached to gorgonians; Figure 5.6L)

A review of the feeding preferences of the members of the *Cyphoma rhomba* Assemblage shows that cnidarivore predators dominate the ecosystem.

Figure 5.6 Mollusks of the *Cyphoma rhomba* Assemblage. A, B = *Cyphoma gibbosum* (Linnaeus, 1758), length 28.7 mm. C = *Cyphoma gibbosum* (Linnaeus, 1758), length 37 mm. Specimen with preserved mantle, showing color pattern. D, E = *Cyphoma mcgintyi* Pilsbry, 1939, length 37.6 mm. F = *Cyphoma mcgintyi* Pilsbry, 1939, length 27.7 mm. Specimen with preserved mantle, showing color pattern. G, H = *Cyphoma rhomba* Cate, 1978, length 16.9 mm. Endemic to the Florida Keys. I = *Cyphoma signatum* Pilsbry and McGinty, 1939, length 33.4 mm. J = *Cyphoma sedlaki* Cate, 1976, length 13.5 mm. Endemic to the Florida Keys. K = *Pseudocyphoma intermedium* (Sowerby II, 1828), length 26.5 mm. L = *Dendostrea frons* (Linnaeus, 1758), length 41 mm. (Note the hook-like shell prongs that are attached to the fragment of gorgonian skeleton.)

Molluscan ecology of the sublittoral exposed hardbottoms

On the shallow subtidal limestone platforms of the Florida Keys, large aggregations of the rock-boring sea urchins *Echinometra lucunter* and *Echinometra viridis* excavate innumerable cavities and pits and create a distinctive, highly eroded seafloor (Figure 5.7). Using their spines as rasps, these specialized urchins carve out rounded microcaves and spend their entire lives within their small, self-created worlds. As the echinoids move around their caves while they feed on algal films, they gradually enlarge their living spaces, and these often anastomose and connect with other microcaves to produce a labyrinth of open cavities. This microcosm of interconnected hidden spaces provides a sheltered habitat for a small but characteristic molluscan assemblage. Because the cerithiid gastropod *Bayericerithium litteratum* (Figure 5.8C, D) is abundant here, we have chosen this small cerith to represent the entire molluscan faunule (the *Bayericerithium litteratum* Assemblage).

Several small sessile bivalves live within the labyrinthine caves of these rock surfaces, all attached by their strong byssal threads. These include the arcids *Arca imbricata* (Figure 5.8H), *Arca zebra* (Figure 5.8L), *Acar domingensis*, *Barbatia cancellaria*, and *Cucullearca candida* (with the last three species also occurring in the *Cerithium lutosum* Assemblage) and the mytilid *Modiolus americanus* (Figure 5.8J). The sea urchin cavities also shelter a small fauna of suspension-feeding sessile gastropods, including the encrusting vermetid *Serpulorbis decussatus* (Figure 5.8G), the hipponicid *Hipponix antiquatus* (Figure 5.8F), and the calyptraeid spiny slipper shell *Bostrycapulus aculeatus* (Figure 5.8I). These sessile forms, along with barnacles, provide the food source for several voracious small carnivorous gastropods, including the fasciolariid *Leucozonia nassa* (Figure 5.8E) and the buccinids *Gemophos tinctus* (Figure 5.8A, B) and *Pisania pusio* (Figure 5.8K). Along many of the intertidal areas of the Florida Keys, the shallow sublittoral *Bayericerithium litteratum* Assemblage intergrades with the littoral *Cerithium lutosum* Assemblage in a broad ecotonal system. The sea urchin minicaves also house a large fauna of polyplacophoran chitons, several of which may prove to be new endemic species.

Biodiversity of the Bayericerithium litteratum *Assemblage*

The malacofauna of the open limestone platforms and sea urchin cavities is relatively impoverished, comprising only a few species. The members of the *Bayericerithium litteratum* Assemblage, listed here by feeding type and ecological niche, are:

1. **HERBIVORES (including algivores)**
   Gastropoda
   Cerithiidae
      Bayericerithium litteratum (Born, 1778) (Figure 5.8C, D)
2. **CARNIVORES (Molluscivores)**
   Gastropoda
   Fasciolariidae
      Leucozonia nassa (Gmelin, 1791) (Figure 5.8E)
   Buccinidae
      Gemophos tinctus (Conrad, 1846) (Figure 5.8A, B)
      Pisania pusio (Linnaeus, 1758) (Figure 5.8K)

Figure 5.7 Close-up of a highly eroded hardbottom seafloor off Missouri Key, Middle Florida Keys, showing the rock-boring sea urchin *Echinometra lucunter*. These echinoids use their spines to excavate deep cavities into the soft limestone, creating a system of interconnected mini-caves and crevices. This type of environment supports the *Bayericerithium litteratum* Assemblage of the Unvegetated Hardbottom Macrohabitat. (Photograph by Ron Bopp.)

Figure 5.8 Mollusks of the *Bayericerithium litteratum* Assemblage. A, B = *Gemophos tinctus* (Conrad, 1846), length 27.7 mm. C, D = *Bayericerithium litteratum* (Born, 1778), length 28 mm. E = *Leucozonia nassa* (Gmelin, 1791), length 37 mm. F = *Hipponix antiquatus* (Linnaeus, 1767), length 14 mm. G = *Serpulorbis decussatus* (Gmelin, 1791), length 28 mm. H = *Arca imbricata* Bruguiere, 1789, length 52 mm. I = *Bostrycapulus aculeatus* (Gmelin, 1791), length 19 mm. J = *Modiolus americanus* (Leach, 1815), length 40 mm. K = *Pisania pusio* (Linnaeus, 1758), length 32 mm. L = *Arca zebra* (Swainson, 1833), length 54 mm.

3. SUSPENSION/FILTER FEEDERS

Gastropoda

Vermetidae

Dendropoma corrodens (d'Orbigny, 1841)

Dendropoma irregulare (d'Orbigny, 1841)

Petaloconchus erectus (Dall, 1888)

Petaloconchus mcgintyi (Olsson and Harbison, 1953)

Petaloconchus varians (d'Orbigny, 1841)

Serpulorbis decussatus (Gmelin, 1791) (Figure 5.8G)

Capulidae

Capulus (Krebsia) incurvatus (Gmelin, 1791)

Hipponicidae

Hipponix antiquatus (Linnaeus, 1767) (Figure 5.8F)

Hipponix subrufus (Lamarck, 1819)

Calyptraeidae

Bostrycapulus aculeatus (Gmelin, 1791) (Figure 5.8I)

Bivalvia

Arcidae

Acar domingensis (Lamarck, 1819)

Arca imbricata Bruguiere, 1789 (Figure 5.8H)

Arca zebra (Swainson, 1833) (Figure 5.8L)

Barbatia cancellaria (Lamarck, 1811)

Cucullearca candida (Helbling, 1779)

Mytilidae

Modiolus americanus (Leach, 1815) (Figure 5.8J)

Pteriidae

Pteria colymbus (Röding, 1798)

A review of the ecological niche preferences of the members of the *Bayericerithium litteratum* Assemblage shows that the suspension feeders dominate the ecosystem.

chapter six

Molluscan faunas of the unvegetated softbottom macrohabitat (nearshore marine regime)

Introduction

Vast areas of the shallow waters off the Florida Keys are composed of thick accumulations of carbonate sand and mud that, characteristically, are devoid of sea grasses and macroalgae. Typically, these submarine "deserts" occur in areas of strong tidal currents, which produce rapidly shifting sediments that are not conducive to plant growth. As is frequently observed, the open unvegetated softbottom areas are scattered among sea grass beds and mangrove forests, usually in tidal channels, and often intergrade into these other environments in broad ecotonal transition zones (Figure 6.1). In the Florida Keys, two main unvegetated softbottom areas occur, both of which are bathymetrically controlled: the Intertidal Open Carbonate Mudflats, which are found primarily along the intertidal zone of the South Florida Bight sides of the islands; and the Sublittoral Open Carbonate Sand Seafloors, which are found primarily within the deeper waters of the Hawk Channel between the Keys and the offshore reef tract. Each of these soft sediment macrohabitats contains its own resident molluscan assemblage, and those of the Hawk Channel habitats are now known to harbor a large number of endemic species.

Molluscan ecology of the intertidal open mudflats

Although not particularly species rich, and encompassing only around 34 species of mollusks, the unvegetated mudflats of the Florida Keys typically house immense aggregations of these gastropods and bivalves. The most obvious inhabitant of the exposed mudflats is the small batillariid gastropod *Batillaria minima* (Figure 6.2C, D), which is so abundant that we have designated it to be the namesake of the entire molluscan faunule (the *Batillaria minima* Assemblage). This molluscan assemblage also contains some of the largest mollusks found within the intertidal zone, including the Left-Handed Whelk, *Sinistrofulgur sinistrum* (Figure 6.3E), the second-largest gastropod found in Florida, and the Campeche Venus, *Mercenaria campechiensis* (Figure 6.2E), one of the largest Florida bivalves. These gigantic mollusks occur along with a host of smaller, shallowly buried bivalves, including the venerids *Chione elevata* (Figure 6.2A) and *Anomalocardia cuneimeris* (Figure 6.3G); arcids such as *Anadara transversa, Anadara floridana, Caloosarca notabilis* (Figure 6.3I), *Lunarca ovalis,* and *Scapharca brasiliana* (Figure 6.3K); the noetiid *Noetia ponderosa* (Figure 6.3J); the donacid *Iphigenia brasiliana* (Figure 6.2B); the multicolored psammobiid *Asaphis deflorata* (Figure 6.2K); the lucinid

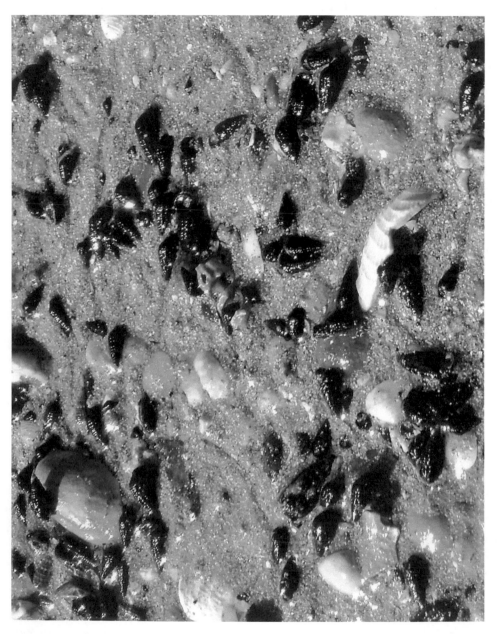

Figure 6.1 Close-up of an aggregation of living *Batillaria minima* (Gmelin, 1791) crawling on an exposed mudflat in Lake Worth, Palm Beach County, Florida. This type of intertidal biotope supports the *Batillaria minima* Assemblage of the Unvegetated Softbottom Macrohabitat. (Photograph by Eddie Matchett.)

Figure 6.2 Mollusks of the *Batillaria minima* Assemblage. A = *Chione elevata* Say, 1822, width 23 mm. B = *Iphigenia brasiliana* (Lamarck, 1818), length 45 mm. C, D = *Batillaria minima* (Gmelin, 1791), length 13 mm. E = *Mercenaria campechiensis* (Gmelin, 1791), width 81 mm. F = *Tagelus plebeius* (Lightfoot, 1786), length 29 mm. G = *Neritina (vitta) virginea* (Linnaeus, 1758), length 12 mm. H = *Cyrtopleura costata* (Linnaeus, 1758), length 131 mm. I = *Neritina (vitta) clenchi* Russel, 1940, length 13 mm. J = *Neritina (vitta) usnea* (Röding, 1798), length 13 mm. K = *Asaphis deflorata* (Linnaeus, 1758), length 42 mm. L = *Tagelus divisus* (Spengler, 1794), length 28 mm.

Figure 6.3 Mollusks of the *Batillaria minima* Assemblage. A, B = *Phrontis polygonatus* (Lamarck, 1822), length 11 mm. C, D = *Phrontis vibex* (Say, 1822), length 10 mm. E = *Sinistrofulgur sinistrum* (Hollister, 1958), length 185 mm. F = *Olivella (Dactylidia) pusilla* (Marrat, 1871), length 7 mm. G = *Anomalocardia cuneimeris* (Conrad, 1846), length 12 mm. H = *Raeta plicatella* (Lamarck, 1818), length 41 mm. I = *Caloosarca notabilis* (Röding, 1798), length 45 mm. J = *Noetia ponderosa* (Say, 1822), length 46 mm. K = *Scapharca brasiliana* (Lamarck, 1819), length 58 mm. L = *Divalinga quadrisulcata* (d'Orbigny, 1842), length 19 mm.

Divalinga quadrisulcata (Figure 6.3L); and the solecurtid razor clams *Tagelus divisus* (Figure 6.2L) and *Tagelus plebeius* (Figure 6.2F). These bivalves are the principal food resource for the large busyconid *Sinistrofulgur sinistrum*, which opens its clam prey by chipping the valve edges with the chisel-like blade of its shell's lip and then inserting its tooth-filled proboscis.

The open intertidal mudflats of the Florida Keys and adjacent areas also harbor an interesting and characteristic fauna of small gastropods, all of which occur in large aggregations of thousands of individuals. The most important of these is a species radiation of estuarine, mud-adapted neritids, including *Neritina (vitta) clenchi* Russel, 1940 (Figure 6.2I), *Neritina (vitta) usnea* (Figure 6.2J), and *Neritina (vitta) virginea* (Figure 6.2G), which feed on algal films on the mud surface; and small scavengers such as the nassariids *Phrontis polygonatus* (Figure 6.3A, B) and *Phrontis vibex* (Figure 6.3C, D) and the olivellid *Olivella (Dactylidia) pusilla* (Figure 6.3F). The mudflats also harbor a small but important fauna of deeply burrowing infaunal bivalves, including the pearly periplomatid *Periploma margaritaceum*; the pholadiform venerid *Petricolaria pholadiformis*; the pholadids *Barnea truncata*, *Pholas campechiensis*, and *Cyrtopleura costata* (Figure 6.2H); and the paper-thin mactrids *Anatina anatina* and *Raeta plicatella* (Figure 6.3H). Several of these mudflat gastropods and bivalves also occur in the *Bulla occidentalis* Assemblage, which has a similar bathymetric range and soft-sediment biotope.

Biodiversity of the Batillaria minima *Assemblage*

The members of the unvegetated intertidal mudflat-dwelling *Batillaria minima* Assemblage are listed here by feeding type and ecological niche preference.

1. HERBIVORES (including algal film feeders)
Gastropoda
Neritidae
> *Neritina (vitta) clenchi* Russel, 1940 (Figure 6.2I)
> *Neritina (vitta) usnea* (Röding, 1798) (Figure 6.2J)
> *Neritina (vitta) virginea* (Linnaeus, 1758) (Figure 6.2G)
Assimineidae
> *Assiminea succinea* (Pfeiffer, 1840)
Batillariidae
> *Batillaria minima* (Gmelin, 1791) (Figure 6.2C, D)
2. CARNIVORES
Gastropoda
2a. MOLLUSCIVORES
Busyconidae
> *Sinistrofulgur sinistrum* (Hollister, 1958) (Figure 6.3E)
Melongenidae
> *Melongena (Rexmela) bicolor* (Say, 1827)
2b. GENERAL CARNIVORES/SCAVENGERS
Nassariidae
> *Phrontis polygonatus* (Lamarck, 1822) (Figure 6.3A, B)
> *Phrontis vibex* (Say, 1822) (Figure 6.3C, D)
Marginellidae
> *Prunum apicinum virgineum* (Jousseaume, 1875)

Cystiscidae-Granulininae
  Granulina hadria (Dall, 1889)
Olivellidae
  Olivella (Dactylidia) pusilla (Marrat, 1871) (Figure 6.3F)
Bullidae
  Bulla occidentalis A. Adams, 1850
3. **SUSPENSION/FILTER FEEDERS**
Bivalvia
Arcidae
  Anadara floridana (Conrad, 1869)
  Anadara transversa (Say, 1822)
  Caloosarca notabilis (Röding, 1798) (Figure 6.3I)
  Lunarca ovalis (Bruguiere, 1789)
  Scapharca brasiliana (Lamarck, 1819) (Figure 6.3K)
Noetiidae
  Noetia ponderosa (Say, 1822) (Figure 6.3J)
Periplomatidae
  Periploma margaritaceum (Lamarck, 1801)
Lucinidae
  Divalinga quadrisulcata (d'Orbigny, 1842) (Figure 6.3L)
  Divaricella dentata (Wood, 1815)
  Phacoides pectinata (Gmelin, 1791)
Veneridae
  Anomalocardia cuneimeris (Conrad, 1846) (Figure 6.3G)
  Chione elevata Say, 1822 (Figure 6.2A)
  Macrocallista nimbosa (Lightfoot, 1786)
  Mercenaria campechiensis (Gmelin, 1791) (Figure 6.2E)
  Petricolaria pholadiformis (Lamarck, 1818)
Tellinidae
  Macoma tenta (Say, 1834)
Donacidae
  Iphigenia brasiliana (Lamarck, 1818) (Figure 6.2B)
Psammobiidae
  Asaphis deflorata (Linnaeus, 1758) (Figure 6.2K)
Solecurtidae
  Tagelus divisus (Spengler, 1794) (Figure 6.2L)
  Tagelus plebeius (Lightfoot, 1786) (Figure 6.2F)
Mactridae
  Anatina anatina (Spengler, 1802)
  Raeta plicatella (Lamarck, 1818) (Figure 6.3H)
Pholadidae
  Barnea truncata (Say, 1822)
  Cyrtopleura costata (Linnaeus, 1758) (Figure 6.2H)
  Pholas campechiensis Gmelin, 1791

A review of the ecological niche preferences of the members of the *Batillaria minima* Assemblage shows that the suspension/filter feeders dominate the ecosystem and make up the greater portion of the biomass.

Molluscan ecology of the sublittoral open carbonate sand seafloors

The unvegetated carbonate sand seafloors, in depths of 1 to 20 m, house the second-largest molluscan assemblage found in the Florida Keys area. With over 152 species of resident mollusks, the malacofauna of these open sand substrates is comparable in species richness only to the molluscan assemblage found on the coral reef tract platforms (with over 192 species). Along the Keys archipelago, most of the seafloors of this type are concentrated within the Hawk Channel and run in a narrow band from Key Biscayne south to the Marquesas Keys. Large patches of this open carbonate sand substrate also occur in the channels between the Lower Keys (Figure 6.5), near the Marquesas Keys and Dry Tortugas, and in Florida Bay along the western sides of the Upper Keys. One of the most characteristic open sand seafloor gastropods found in these areas (Figures 6.6) is the Moon Snail *Polinices lacteus* (Figure 6.4G), which is restricted to clean carbonate sand substrates and is here chosen to represent the entire macrohabitat (the *Polinices lacteus* Assemblage).

The sand seafloors of the Hawk Channel contain one of the richest macrobivalve faunas found in the tropical western Atlantic, with over 87 species in 48 genera and belonging to 14 different families (see species list at the end of this section). Of these, the family Tellinidae is the largest, containing at least 32 species and including conspicuous forms such as *Laciolina magna* (Figure 6.7I) and *Eurytellina lineata* (Figure 6.7K). Second in species richness to the Tellinidae are the Venus Clams of the family Veneridae, with over 18 species. The largest of the venerids, and also one of the most conspicuous species, is the Checker-Board Venus, *Callista maculata* (Figure 6.7H), which occurs in the deeper parts of the Hawk Channel. This very large bivalve fauna, which also includes the pectinid scallops *Argopecten gibbus* (6.7J) and *Nodipecten fragosus* (Figure 6.7L) and the semelid *Semele purpurascens* (Figure 6.7M), serves as the principal food resource for an exceptionally large and distinctive fauna of drilling molluscivorous naticid gastropods, including the aforementioned *Polinices lacteus* (Figure 6.4G), the small *Polinices uberinus* (Figure 6.6H), the highly colored *Naticarius canrena* (Figure 6.4F), *Stigmaulax sulcatus* (Figure 6.4M), and the flattened *Sinum maculosum* (Figure 6.8J).

A large fauna of general carnivores and scavengers also occurs on the open carbonate sand substrate areas, some of which include the nassariids *Uzita alba* (Figure 6.4I, J), *Uzita paucicostata* (Figure 6.4K, L), and *Uzita antillarum*; the olivellids *Jaspidella jaspidea* (Figure 6.7A), *Olivella (Dactylidia) dealbata* (Figure 6.7B), *Olivella floralia* (Figure 6.7C), and *Olivella (Dactylidia) mutica* (Figure 6.7E); the olivid *Americoliva bollingi* (and its color form *pattersoni*; Figure 6.8A–E); the volutid *Scaphella junonia elizabethae* (Figure 6.8G, H); the marginellid *Hyalina avenacea* (Figure 6.8K); and the cystiscids *Gibberula lavalleeana* and *Pugnus serrei*. The coarse coralline sand also supports a rich fauna of infaunal polychaete worms, which are the principal prey items of a diverse vermivorous gastropod fauna, some of which include the terebrids *Hastula hastata* (Figure 6.7D), *Strioterebrum protextum* (Figure 6.7F), *Myurellina taurina* (Figure 6.8F), and *Strioterebrum dislocatum onslowensis* (Figure 6.7G; a subspecies, or possibly full species, that is found in clean carbonate sediments near coral bioherms off North Carolina and Florida); the conids *Dauciconus daucus* (Figure 6.9A), *Attenuiconus attenuatus* (Figure 6.9B), *Tuckericonus flavescens caribbaeus* (Figure 6.9C, D), and *Lindaconus atlanticus* (Figure 6.9E, F); and the conilithids *Jaspidiconus fluviamaris* (Figure 6.9G, H), *Jaspidiconus vanhyningi* (Figure 6.9I, J), and *Jaspidiconus pfluegeri* (Figure 6.9K–M). All of these vermivores bury themselves in the sand during the day and become active at night, when they search for their free-crawling sea worm prey. Another

Figure 6.4 Gastropods of the *Polinices lacteus* Assemblage. A, B = *Semicassis granulata* (Born, 1778), length 64 mm. C, D = *Semicassis cicatricosa* (Gmelin, 1791), length 32 mm. E = *Casmaria atlantica* Clench, 1944, length 26 mm. F = *Naticarius canrena* (Linnaeus, 1758), width 28 mm. G = *Polinices lacteus* (Guilding, 1834), length 24 mm. H = *Polinices uberinus* (d'Orbigny, 1842), length 14 mm. I, J = *Uzita alba* (Say, 1826), length 7 mm. K, L = *Uzita paucicostata* (Marrat, 1877), length 8 mm. M = *Stigmaulax sulcatus* (Born, 1778), length 18 mm.

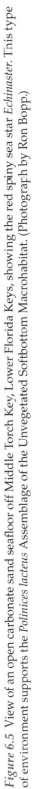

Figure 6.5 View of an open carbonate sand seafloor off Middle Torch Key, Lower Florida Keys, showing the red spiny sea star *Echinaster*. This type of environment supports the *Polinices lacteus* Assemblage of the Unvegetated Softbottom Macrohabitat. (Photograph by Ron Bopp.)

Figure 6.6 Gastropods of the *Polinices lacteus* Assemblage. A, B = *Macrostrombus costatus* (Gmelin, 1791), length 114 mm. C, D = *Strombus pugilis* Linnaeus, 1758, length 83 mm. E, F = *Strombus alatus* Gmelin, 1791, length 99 mm. G, H = *Cassis spinella* Clench, 1944, length 263 mm. I, J = *Cassis madagascariensis* Lamarck, 1822, length 272 mm. K, L = *Cassis tuberosa* (Linnaeus, 1758), length 173 mm.

carnivorous gastropod, the cancellariid *Cancellaria adelae* (Figure 6.8I), possibly feeds on sleeping stingrays, inserting its proboscis and needle-like teeth into the gill chambers and sucking the blood from the gill filaments.

The open sand seafloors also house a large fauna of irregular echinoid echinoderms (Heart Urchins and Cake Urchins), including large species with thin tests such as *Plagiobrissus grandis* and *Meoma ventricosa*, and several species of flattened sand dollars, including *Mellita quinquesperforata* and *Mellita sexiesperforata*. The large, inflated *Meoma* and *Plagiobrissus* echinoids serve as the principal food source for three large echinoid-feeding cassid gastropods, including *Cassis spinella* (a Carolinian Molluscan Province endemic species; Figure 6.6G, H), *Cassis madagascariensis* (a Caribbean Molluscan Province species that ranges to Florida; Figure 6.6I, J), and *Cassis tuberosa* (a widespread western Atlantic species; Figure 6.6K, L). These large species occur together with three smaller cassids: *Semicassis granulata* (Figure 6.4A, B), *Semicassis cicatricosa* (Figure 6.4C, D), and *Casmaria atlantica* (Figure 6.4E), all of which feed on the abundant *Mellita* sand dollars. In a similar fashion to the naticids, these cassids use their radular teeth to rasp holes in the thin tests of the echinoids and tear into the internal organs. The large tonnid gastropods *Tonna galea* and *Tonna pennata* (Figure 6.8L) also feed on sand-dwelling echinoderms, in this case several species of infaunal holothurians sea cucumbers.

Interestingly, the only prominent algivores found on this open carbonate sand macrohabitat are strombid gastropods. Two of these algivores, the small Florida Fighting Conch (*Strombus alatus*, a Carolinian Molluscan Province endemic species; Figure 6.6E, F) and the Caribbean Fighting Conch (*Strombus pugilis*, a Caribbean Molluscan Province species that ranges into southern Florida; Figure 6.6C, D), live in huge shoal-like aggregations, often numbering in the thousands of individuals. These small fighting conchs crawl across the open seafloor, utilizing their long, elephantine proboscises to feed on the algal films that grow on the sand surface. Occurring with the small *Strombus* species, but never as abundantly, is the Milk Conch, *Macrostrombus costatus* (Figure 6.6A, B). This conch with a much larger shell often feeds on blue-green algal films and can accumulate toxins (especially ciguatoxins) within its tissues, producing ciguatera poisoning in any human who consumes it.

Biodiversity of the Polinices lacteus *Assemblage*

The open sand substrates of the Hawk Channel and adjacent areas house a number of important endemic Florida Keys and southeastern Florida mollusks. Some of these include gastropods such as the cancellariid *Cancellaria adelae*; the shallow-water volutid *Scaphella junonia elizabethae*; the conid *Tuckericonus flavescens caribbaeus* (which ranges from Palm Beach to the Dry Tortugas); and the conilithids *Jaspidiconus fluviamaris* (which ranges from Palm Beach to the Dry Tortugas), *Jaspidiconus pfluegeri* (which ranges from eastern Florida to the Middle Florida Keys; several specimens were recently collected on open sand areas off Missouri Key), and *Jaspidiconus vanhyningi* (which ranges from Palm Beach to the Upper Keys); and the bivalve *Tivela floridana* (also found along the Palm Beach coast). For illustrations of the bivalves listed here (under Suspension/Filter Feeders), see Mikkelsen and Bieler (2008). The members of the sublittoral open carbonate sand-dwelling *Polinices lacteus* Assemblage are listed here according to feeding type and ecological niche preferences.

1. **HERBIVORES (including algal film feeders)**
   Gastropoda
   Strombidae
   Macrostrombus costatus (Gmelin, 1791) (Figure 6.6A, B)
   Strombus alatus Gmelin, 1791 (Figure 6.6E, F)
   Strombus pugilis Linnaeus, 1758 (Figure 6.6C, D)
   Architectonicidae (also feeds on foraminifera)
   Architectonica nobilis Röding, 1798
2. **CARNIVORES**
   Gastropoda
2a. **GENERAL CARNIVORES/SCAVENGERS**
   Eulimidae (only large species listed here; possibly feed on sand-dwelling zoantherians)
   Niso aeglees Bush, 1885
   Niso hendersoni Bartsch, 1953
   Buccinidae
   Antillophos candeanus (d'Orbigny, 1842) (in deeper areas of the Hawk Channel)
   Nassariidae
   Uzita alba (Say, 1826) (Figure 6.4I, J)
   Uzita antillarum (d'Orbigny, 1842)
   Uzita paucicostata (Marrat, 1877) (Figure 6.4K, L)
   Olivellidae
   Jaspidella blanesi (Ford, 1898)
   Jaspidella jaspidea (Gmelin, 1791) (Figure 6.7A)
   Olivella floralia (Duclos, 1853) (Figure 6.7C)
   Olivella lactea Marrat, 1871 (= *adelae* Olsson, 1956)
   Olivella nivea (Gmelin, 1791)
   Olivella (Dactylidia) dealbata (Reeve, 1850) (Figure 6.7B)
   Olivella (Dactylidia) mutica (Say, 1822) (Figure 6.7E)
   Olividae
   Americoliva bollingi (Clench, 1934) (Figure 6.8A–D)
   Volutidae-Lyriinae
   Enaeta cylleniformis (Sowerby I, 1844) (This small volute, which is common in the Bahamas, has been reported from northern Biscayne Bay by several collectors and workers; apparently known from only a few specimens, its presence in the Florida Keys Reef Tract area has yet to be firmly established.)
   Volutidae-Scaphellinae
   Scaphella junonia elizabethae Petuch and Sargent, 2011 (Figure 6.8G, H)
   Marginellidae
   Hyalina avenacea (Deshayes, 1844) (Figure 6.8K)
   Prunum bellum (Conrad, 1868)
   Volvarina albolineata (d'Orbigny, 1842)
   Cystiscidae (Persiculinae and Granulininae)
   Gibberula lavalleeana (d'Orbigny, 1842)
   Pugnus serrei (Bavay, 1911)
   Pyramidellidae (large species only)
   Pyramidella dolabrata (Linnaeus, 1758)
   Cylichnidae
   Cylichnella bidentata (d'Orbigny, 1841)

Haminoeidae
Atys macandrewi E.A. Smith, 1872
Acteocinidae
Acteocina candei (d'Orbigny, 1842)
Acteocina inconspicua Olsson and McGinty, 1958
Acteocina recta (d'Orbigny, 1841)
Acteonidae
Acteon candens Rehder, 1939
Acteon punctostriata (C.B. Adams, 1840)
2b. SUCTORIAL FEEDERS (possibly on stingrays)
Cancellariidae
Cancellaria adelae Pilsbry, 1940 (Figure 6.8I)
2c. ECHINODERM FEEDERS
Cassidae
Cassis madagascariensis Lamarck, 1822 (Figure 6.6I, J)
Cassis spinella Clench, 1944 (Figure 6.6G, H)
Cassis tuberosa (Linnaeus, 1758) (Figure 6.6K, L)
Casmaria atlantica Clench, 1944 (Figure 6.4E)
Semicassis cicatricosa (Gmelin, 1791) (Figure 6.4C, D)
Semicassis granulata (Born, 1778) (Figure 6.4A, B)
Tonnidae
Tonna galea (Linnaeus, 1758)
Tonna pennata (Mörch, 1852) (Figure 6.8L)
2d. MOLLUSCIVORES (drilling predators)
Naticidae
Naticarius canrena (Linnaeus, 1758) (Figure 6.4F)
Naticarius tedbayeri (Rehder, 1986)
Natica livida Pfeiffer, 1840
Natica marochiensis (Gmelin, 1791)
Polinices hepaticus (Röding, 1798)
Polinices lacteus (Guilding, 1834) (Figure 6.4G)
Polinices uberinus (d'Orbigny, 1842) (Figure 6.4H)
Sigatica semisulcata (Gray, 1839)
Sinum maculosum (Say, 1831) (Figure 6.8J)
Sinum perspectivum (Say, 1831)
Stigmaulax sulcatus (Born, 1778) (Figure 6.4M)
Tectonatica pusilla (Say, 1822)
2e. VERMIVORES
Conidae-Puncticulinae
Attenuiconus attenuatus (Reeve, 1844) (Figure 6.9B)
Dauciconus daucus (Hwass, 1792) (Figure 6.9A)
Lindaconus atlanticus (Clench, 1942) (Figure 6.9E, F)
Tuckericonus flavescens caribbaeus (Clench, 1942) (Figure 6.9C, D)
Conilithidae-Conilithinae
Jaspidiconus fluviamaris Petuch and Sargent, 2011 (Figure 6.9G, H)
Jaspidiconus pfluegeri Petuch, 2004 (Figure 6.9K–M)
Jaspidiconus vanhyningi (Rehder, 1944) (Figure 6.9I, J)

Terebridae
  Hastula cinerea (Born, 1778) (Figure 6.8E)
  Hastula hastata (Gmelin, 1791) (Figure 6.7D)
  Hastula maryleeae Burch, 1965
  Hastula salleana (Deshayes, 1859) (Figure 6.8M)
  Myurella taurina (Lightfoot, 1786) (Figure 6.8F)
  Strioterebrum concavum (Say, 1827)
  Strioterebrum dislocatum onslowensis (Petuch, 1974) (Figure 6.7G)
  Strioterebrum glossema (Schwengel, 1940)
  Strioterebrum protextum (Conrad, 1845) (Figure 6.7F)
Drilliidae
  Neodrillia moseri (Dall, 1889)
  Neodrillia wolfei (Tippett, 1995)
Clathurellidae
  Nannodiella melanitica (Bush, 1885)
  Nannodiella oxia (Bush, 1885)
  Nannodiella pauca Fargo, 1953
Raphitomidae
  Glyphoturris quadrata (Reeve, 1845)
3. SUSPENSION/FILTER FEEDERS
Bivalvia
Nuculidae
  Nucula proxima Say, 1822
Nuculanidae
  Nuculana concentrica (Say, 1824)
Glycymeridae
  Glycymeris decussata (Linnaeus, 1758)
  Glycymeris spectralis Nicol, 1952
  Glycymeris undata (Linnaeus, 1758)
Pectinidae
  Argopecten gibbus (Linnaeus, 1758) (Figure 6.7J)
  Euvola ziczac (Linnaeus, 1758)
  Nodipecten fragosus (Conrad, 1849) (Figure 6.7L)
Crassitellidae
  Crassinella martinicensis (d'Orbigny, 1845)
  Eucrassatella speciosa (A. Adams, 1852)

Figure 6.7 Mollusks of the *Polinices lacteus* Assemblage. A = *Jaspidella jaspidea* (Gmelin, 1791), length 15 mm. B = *Olivella (Dactylidia) dealbata* (Reeve, 1850), length 6 mm. C = *Olivella floralia* (Duclos, 1853), length 8 mm. D = *Hastula hastata* (Gmelin, 1791), length 27 mm. E = *Olivella (Dactylidia) mutica* (Say, 1822), length 6 mm. F = *Strioterebrum protextum* (Conrad, 1845), length 28 mm. G = *Strioterebrum dislocatum onslowensis* (Petuch, 1974), length 32 mm. This subspecies (or possible full species) is confined to clean carbonate sand areas near coral bioherms off North Carolina, Georgia, and the Florida Keys. Although originally described from the coral bioherms in Onslow Bay, North Carolina, this finely ribbed specimen was found on an open carbonate sand area off Missouri Key, Middle Florida Keys (see Petuch, 1974). H = *Callista maculata* (Linnaeus, 1758), length 54 mm. I = *Laciolina magna* (Spengler, 1798), length 71 mm. J = *Argopecten gibbus* (Linnaeus, 1758), width 30 mm. K = *Eurytellina lineata* (Turton, 1819), length 28 mm. L = *Nodipecten fragosus* (Conrad, 1849), width 87 mm. M = *Semele purpurascens* (Gmelin, 1791), width 28 mm.

Lucinidae
Divalinga dentata (Wood, 1815)
Divalinga quadrisulcata (d'Orbigny, 1845)
Carditidae
Pleuromeris tridentatus (Say, 1826)
Chamidae
Arcinella cornuta Conrad, 1866
Cardiidae
Acrosterigma magnum (Linnaeus, 1758)
Americardia guppyi (Thiele, 1910) (Note: Lee and Huber, 2012, have pointed out
that Mikkelsen and Bieler, 2008, incorrectly placed *Americardia guppyi* and *A. media* in *Ctenocardia*, which is a South Pacific genus.)
Americardia media (Linnaeus, 1758)
Dallocardia muricata (Linnaeus, 1758)
Papyridea semisulcata (Gray, 1825)
Papyridea soleniformis (Bruguiere, 1789)
Trigonocardia antillarum (d'Orbigny, 1842)
Veneridae
Callista maculata (Linnaeus, 1758) (Figure 6.7H)
Chione cancellata (Linnaeus, 1767) (rare in Florida; deeper water only)
Chione elevata (Say, 1822)
Chionopsis intapurpurea (Conrad, 1849)
Chionopsis pubera (Bory de Saint-Vincent, 1827)
Cooperella atlantica Rehder, 1943
Gemma gemma (Totten, 1834)
Gouldia cerina (C.B. Adams, 1845)
Parastarte triquetra (Conrad, 1846)
Pitarenus albidus (Gmelin, 1791)
Pitarenus cordatus (Schwengel, 1951)
Pitar circinatus (Born, 1778)
Pitar dione (Linnaeus, 1758)

Figure 6.8 Gastropods of the *Polinices lacteus* Assemblage. A, B = *Americoliva bollingi* (Clench, 1943), length 49 mm. This common endemic eastern Florida and Florida Keys olive has often been mis-identified as *Oliva bifasciata* Küster, 1878, or has been considered a subspecies of it (as *Oliva bifasciata bollingi*). The true *Americoliva bifasciata* is actually a larger, more inflated species that is confined to the West Indian Arc and Caribbean Basin and is not found in the Florida Keys area. (For illustrations of true West Indian *bifasciata*, see Petuch and Sargent, 1986, Plate 21, Figures 9–12, 15.) C = *Americoliva bollingi* (Clench, 1943), banded color form, length 46 mm. D = *Americoliva bollingi* (Clench, 1943), melanistic color form *pattersoni* Clench, 1934, length 45 mm. E = *Hastula cinerea* (Born, 1778), length 28 mm. F = *Myurella taurina* (Lightfoot, 1786), length 113 mm. G, H = *Scaphella junonia elizabethae* Petuch and Sargent, 2011, length 65 mm. This Florida Keys and Dry Tortugas endemic subspecies differs from the nominate subspecies in having a smaller shell with a proportionally lower spire and in having fewer, smaller, more rounded spots (a color pattern similar to the subspecies *butleri* from Yucatan, Mexico). Although the other three subspecies of *S. junonia* live in deeper-water areas, *S. junonia elizabethae* lives in shallow-water sandy areas near coral reefs, often in depths of as little as 3 m (as on Pickles Reef off Plantation Key, where the figured specimen was collected). I = *Cancellaria adelae* Pilsbry, 1940, length 39 mm. J = *Sinum maculosum* (Say, 1831), width 28 mm. K = *Hyalina avenacea* (Deshayes, 1844), length 10 mm. L = *Tonna pennata* (Mörch, 1852), length 59 mm. (= *T. maculosa*). M = *Hastula salleana* (Deshayes, 1859)

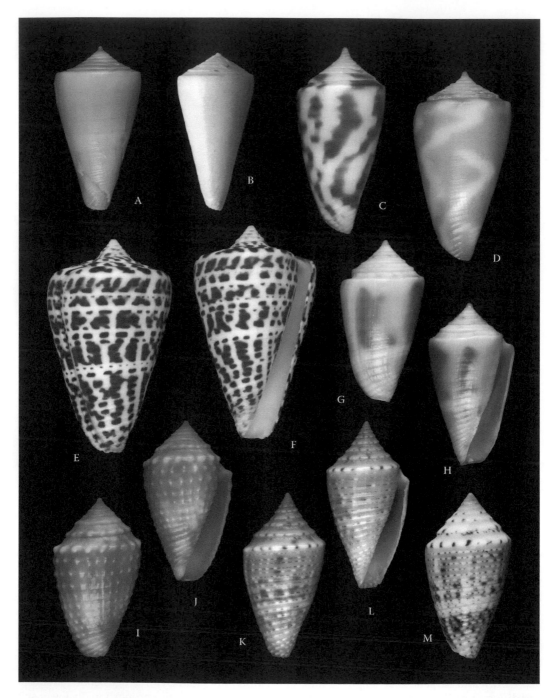

Figure 6.9 Cone Shells of the *Polinices lacteus* Assemblage. A = *Dauciconus daucus* (Hwass, 1792), length 28 mm. B = *Attenuiconus attenuatus* (Reeve, 1844), length 22 mm. C = *Tuckericonus flavescens caribbaeus* (Clench, 1942), length 15 mm. D = *Tuckericonus flavescens caribbaeus* (Clench, 1942), length 15.5 mm. E, F = *Lindaconus atlanticus* (Clench, 1942), length 51 mm. G, H = *Jaspidiconus fluviamaris* Petuch and Sargent, 2011, length 16 mm. I, J = *Jaspidiconus vanhyningi* (Rehder, 1944), length 16 mm. K, L = *Jaspidiconus pfluegeri* Petuch, 2004, length 17 mm. M = *Jaspidiconus pfluegeri* Petuch, 2004, length 17 mm, from off Missouri Key, where it lives sympatrically with *Jaspidiconus pealii*.

  Timoclea grus (Holmes, 1858)
  Tivela abaconis Dall, 1902
  Tivela floridana Rehder, 1939
  Tivela mactroides (Born, 1778)
  Transenella conradiana (Dall, 1884)
  Transenella cubaniana (d'Orbigny, 1853)
Tellinidae
  Acorylus gouldi (Hanley, 1846)
  Angulus agilis (Stimpson, 1857)
  Angulus merus (Say, 1834)
  Angulus paramerus (Boss, 1964)
  Angulus probinus (Boss, 1964)
  Angulus sybariticus (Dall, 1881)
  Angulus texanus (Dall, 1900)
  Angulus versicolor (DeKay, 1843)
  Elliptotellina americana (Dall, 1900)
  Eurytellina angulosa (Gmelin, 1791)
  Eurytellina lineata (Turton, 1819) (Figure 6.7K)
  Eurytellina nitens (C.B. Adams, 1845)
  Eurytellina punicea (Born, 1778)
  Laciolina laevigata (Linnaeus, 1758)
  Laciolina magna (Spengler, 1798) (Figure 6.7I)
  Leporimetis intastriata (Say, 1826)
  Macoma constricta (Bruguiere, 1792)
  Macoma extenuata Dall, 1900
  Macoma limula Dall, 1895
  Macoma pseudomera Dall and Simpson, 1901
  Macoma tageliformis Dall, 1900
  Merisca aequistriata (Say, 1824)
  Merisca crystallina (Spengler, 1798)
  Merisca martinicensis (d'Orbigny, 1853)
  Phyllodina squamifera (Deshayes, 1855)
  Scissula consobrina (d'Orbigny, 1853)
  Scissula iris (Say, 1822)
  Strigilla carnaria (Linnaeus, 1758)
  Strigilla mirabilis (Philippi, 1841)
  Strigilla pisiformis (Linnaeus, 1758)
  Tellina radiata (Linnaeus, 1758)
Psammobiidae
  Gari circe (Mörch, 1876)
Semelidae
  Abra americana Verrill and Bush, 1898
  Abra lioica (Dall, 1881)
  Cumingia coarctata Sowerby I, 1833
  Cumingia vanhyningi Rehder, 1939
  Ervilia concentrica (Holmes, 1858)
  Ervilia nitens (Montagu, 1808)
  Ervilia subcancellata E.A. Smith, 1885
  Semele proficua (Pulteney, 1799)

      Semele purpurascens (Gmelin, 1791) (Figure 6.7M)
      Semelina nuculoides (Conrad, 1841)
Solecurtidae
      Solecurtus cumingianus (Dunker, 1861)
Pharidae
      Ensis minor Dall, 1900
Corbulidae
      Caryocorbula caribaea (d'Orbigny, 1853)
      Caryocorbula chittyana (C.B. Adams, 1852)
      Caryocorbula dietziana (C.B. Adams, 1852)
      Juliacorbula aequivalvis (Philippi, 1836)
      Variacorbula limatula (Conrad, 1846)
      Variacorbula philippii (E.A. Smith, 1845)
  4. CHEMOSYNTHETIC FEEDERS (containing symbiotic chemoautotrophic bacteria)
Solemyidae
      Solemya velum Say, 1822

A review of the ecological niche preferences of the members of the *Polinices lacteus* Assemblage shows that the suspension/filter-feeding bivalves dominate the ecosystem.

Mollusks of the mixed carbonate sand and coral rubble seafloors

Immediately adjacent to the living reef corals of the Florida Reef Tract are large rubble piles that are composed of broken fragments of dead coral. These coral shards often mix with the coarse clean carbonate sand to form a wide transitional biotope between the open sand seafloors (the *Polinices lacteus* Assemblage) and the living coral reef platforms (the *Stephanoconus regius* Assemblage; see Chapter 7). The coral fragments act as a substrate for the attachment of small brown algae, such as *Dictyota*, which form the basis of a small subsidiary ecotonal community. One of the more prominent and conspicuous gastropods found in this transitional deme is the large Queen Conch, *Eustrombus gigas* (Figure 6.10H, I), which continuously crops the algal growth in a bovine fashion. Although occurring on a number of different biotopes, including Turtle Grass beds and open sand substrates, *Eustrombus gigas* prefers the mixed-coral rubble zone and can be found there in large aggregations. These algivorous giants cooccur in the brown algae beds with a number of other smaller herbivorous gastropods, including the turbinids *Lithopoma caelata* and *Lithopoma tuber*, the amphiatlantic cerithiid *Cerithium guinaicum*, and the xenophorid Carrier Shell *Xenophora conchyliophora*. In deeper areas, in depths ranging from 10 to 25 m, beds of Star Grass (*Halophila engelmanni*) often grow on these mixed sand and coral rubble substrates. North of the Dry Tortugas, in the crystal clear waters of the Tortugas Gyre, this deep-dwelling marine angiosperm has been found growing in over 30 m of water, and its beds may house a completely new and unexplored malacofauna.

A small but interesting fauna of specialized carnivorous gastropods also occurs in the coral rubble zone, including a radiation of epitoniid cnidarian predators. These delicate and beautiful gastropods feed primarily on large sand-dwelling zoantherian sea anemones, from which they suck the body fluids through their elongated, needle-tipped proboscises. At least four species of epitoniids are found in this area, with *Gyroscala lamellosa* (Figure 6.10J) the largest and most commonly encountered. Of special interest in the coral rubble zone is the vermivorous cone shell *Gradiconus anabathrum tranthami* (Figure 6.10A–D), which is ecologically confined to this transitional ecotone. This Florida Keys endemic conid is found in coral rubble areas near living reefs from Key Biscayne south to the Dry Tortugas. Throughout its range, *Gradiconus anabathrum tranthami* varies greatly in color, being white with orange patches (Figure 6.10A, B; similar to the juvenile holotype from Pickles Reef), tan with rows of darker spots, or pink with tan spots (Figure 6.10C, D). In the Dry Tortugas area, all of these color forms occur together, and this local population is the most colorful known for the species.

The small and delicate muricid *Vokesimurex rubidus* (Figure 6.10E, F) is frequently encountered in the coral rubble zone, where it is often found nestled among the algal clumps. Like the Queen Conch, this eurytopic muricid occurs on a number of different biotopes, including Turtle Grass beds, but appears to prefer a mixed sand and rubble habitat. Here, it feeds on small gastropods and interstitial bivalves, drilling through their shells with its radular teeth. *Vokesimurex rubidus* lives sympatrically with several other rare and seldom-seen carnivorous gastropods, including the Florida Keys endemic mitrid *Dibaphimitra florida* (Figure 6.10G) and the suctorial-feeding cancellariid *Bivetopsia rugosum*.

The members of the mixed coral rubble and carbonate sand habitat are arranged here by feeding type and ecological niche preference. Some of the most frequently encountered species include the following:

1. HERBIVORES (including algivores)
Gastropoda
Turbinidae
  Lithopoma caelata (Gmelin, 1791)
  Lithopoma tuber (Linnaeus, 1767)
Cerithiidae
  Cerithium guinaicum Philippi, 1849
Strombidae
  Eustrombus gigas (Linnaeus, 1758) (Figure 6.10H, I)
Xenophoridae
  Xenophora conchyliophora (Born, 1780)
2. CARNIVORES
Gastropoda
2a. GENERAL CARNIVORES/SCAVENGERS
Mitridae
  Dibaphimitra florida (Gould, 1856) (Figure 6.10G)
2b. MOLLUSCIVORES (drilling predators)
Muricidae-Muricinae
  Vokesimurex rubidus (F.C. Baker, 1897) (Figure 6.10E, F; also found in Turtle Grass beds)
2c. CNIDARIAN FEEDERS (suctorial on zoantherians)
Epitoniidae
  Epitonium albidum (d'Orbigny, 1842)
  Epitonium foliaceicostum (d'Orbigny, 1842)
  Gyroscala lamellosa (Lamarck, 1822) (Figure 6.10J)
  Pictoscala rupicola (Kurtz, 1860)

2d. VERMIVORES
Conidae-Puncticulinae
Gradiconus anabathrum tranthami (Petuch, 1995) (Figure 6.10A–D)
Conilithidae-Conilithinae
Jaspidiconus fluviamaris Petuch and Sargent, 2011
Jaspidiconus pfluegeri Petuch, 2004
Jaspidiconus vanhyningi (Rehder, 1944)
2e. PROBLEMATICAL (suctorial on fish or polychaete worms?)
Cancellariidae
Bivetopsia rugosum (Lamarck, 1822)

Mollusks of the Palm Beach coastal lagoons

The coastline of Palm Beach County marks the northern terminus of the high tropical environments that characterize the coral coast and Florida Keys area. Prominent intertidal and shallow subtidal Caribbean Molluscan Province index species, such as the gastropods *Cenchritis muricatus*, *Plicopurpura patula*, *Stramonita rustica*, and *Strombus pugilis*, range only to Palm Beach, demonstrating that the oceanographic conditions undergo a dramatic change just north of this area. From Miami, Dade County, to Jupiter, Palm Beach County, the warm Gulf Stream current hugs the shoreline, often meandering to within less than 1 km of the coast. This high tropical water mass regularly spins off small current loops, which can actually enter some of the coastal embayments, bringing Caribbean-type conditions to the shorelines. North of Jupiter, the Gulf Stream curves away from the shoreline, carrying the tropical water far offshore and causing the central Florida coastal water conditions to more closely resemble those of cooler areas such as Georgia and South Carolina. This is particularly noticeable during the winter months, when water temperatures along the Central Florida coast (Fort Pierce northward) can dip below minimum tropical temperatures (below 68°F), while those of Palm Beach County remain far warmer. This oceanographic transitional area is referred to as the Palm Beach Provinciatone, where cold-tolerant mollusks from the Georgian Molluscan Subprovince (to the north) coexist with both tropical mollusks from the Floridian Molluscan Subprovince (to the south) and mollusks that are endemic to the transitional area (provinciatonal endemics; see Petuch, 2013: 40–44).

North of Broward County, the coastline is characterized by a series of large saltwater lagoons, tidal creeks, and estuaries, the largest of these being the Lake Worth lagoon system

Figure 6.10 Gastropods of the Mixed Carbonate Sand and Coral Rubble Substrates. A, B = *Gradiconus anabathrum tranthami* (Petuch, 1995), length 26 mm, from Pickles Reef, Plantation Key (topotype specimen from the type locality; similar in shape and color pattern to white specimens of *anabathrum tranthami* from the Dry Tortugas). The holotype from Pickles Reef, although similar in color to this specimen, was a juvenile individual and had a more pyriform shell outline (a character that disappears as the animal matures). C, D = *Gradiconus anabathrum tranthami* (Petuch, 1995), length 31 mm. This is the pink color form from the Dry Tortugas that was named *tortuganus* Petuch and Sargent, 2011 (*antoni* Cargile, 2011, is another name for this distinctive color variety). Brown, pink, yellow, and white color forms all occur together on the reefs off the Dry Tortugas; white and yellow color forms predominate along the northern and central Keys. E, F = *Vokesimurex rubidus* (F.C. Baker, 1897), length 33 mm. Compare with *Vokesimurex rubidus* form *marcoensis* (Figure 5.5D, E, Chapter 5), which has a higher spire, more sloping whorls, a more rounded shoulder, and a purple aperture. G = *Dibaphimitra florida* (Gould, 1856), length 47 mm. H, I = *Eustrombus gigas* (Linnaeus, 1758), length 240 mm. J = *Gyroscala lamellosa* (Lamarck, 1822), length 23 mm.

of central and northern Palm Beach County (Figure 6.11). This extensive mangrove-lined coastal lagoonal complex is the remnant of a series of large estuarine systems that occurred along the southeastern Florida coast during the late Pleistocene (Petuch and Roberts, 2007: 173–174). Several of these ancient estuaries, such as the Seminole Lagoon System (Sangamonian Stage of the Tarantian Pleistocene; see Petuch and Roberts, 2007: 174–175), evolved highly endemic molluscan faunas, and many of these local endemics are still living in the Recent coastal lagoon systems. Some classic examples of Pleistocene Seminole Lagoon species that are extant in Lake Worth include gastropods such as the melongenid *Melongena (Rexmela) corona winnerae* (Figure 6.12J) and the modulids *Modulus pacei* (Figure 6.12E, F) and *Modulus calusa foxhalli* new subspecies (Figure 6.13H) and bivalves such as the semelid *Semele donovani* (Figure 6.13B). These lagoonal endemics are also known as late Pleistocene fossils, and their shells are frequently encountered in the highly fossiliferous beds of the Ft. Thompson Formation (Coffee Mill Hammock Member) [see Abbott, 1974: 513, for *Semele donovani* and Petuch, 2004: 252–253; Petuch and Roberts, 2007: 183–186 for *Melongena (Rexmela) corona winnerae* and *Modulus pacei*]. The shallow intertidal muddy-sand areas of the Palm Beach coastal lagoons also house a distinctive subspecies of the Texan and Gulf of Mexico moon snail *Neverita delessertiana*. This endemic southeastern Floridian subspecies, *Neverita delessertiana patriceae* (Figure 6.12G–I; described in the Systematic Appendix at the end of this book), differs from its Texan and Gulf of Mexico relative in having a much higher spire, distinct conical shape, flattened shell base, smaller umbilical callus, and larger, deeper, and more open umbilical area. Of the resident naticid moon snails in Lake Worth Lagoon, *patriceae* is the largest taxon and is one of the largest drilling molluscivorous carnivores in the Palm Beach Provinciatone region (see Petuch, 2013: 40–44).

The very large "living fossil" subspecies of *Melongena (Rexmela) corona*, described as *corona winnerae*, ranges only from the St. Lucie River estuary of Martin County to Lake Worth and is the farthest-south member of the *Melongena corona* species complex living along southeastern Florida. The coastline between Lake Worth and northern Biscayne Bay is devoid of any melongenids, and the family reappears in Biscayne Bay as the closely related, but different, *Melongena bicolor*. Other closely related *corona* subspecies reappear in the Ten Thousand Islands and range northward to Mobile Bay, demonstrating that the species is bifurcated into two separate complexes: one ranging from St. Augustine to Lake Worth on the east and one from Mobile Bay to the Ten Thousand Islands on the west. The Keys endemic *Melongena (Rexmela) bicolor* occupies the area between the two complexes of *Melongena corona* subspecies, and the Florida Keys carbonate environments appear to act as a biogeographical and genetic barrier.

The shorelines of Lake Worth characteristically contain small outcrops of highly weathered coquina limestone of late Pleistocene age (the Anastasia Formation), and these are typically randomly scattered among the coastal mangrove forests. The hard surfaces of the Anastasia outcrops provide a substrate for the attachment of three species of oysters: *Crassostrea virginica*, *Crassostrea rhizophorae*, and *Ostreola equestris*. These sessile bivalves often accumulate in immense masses. The oyster clumps and exposed rock surfaces house

Figure 6.11 View of a tidal creek at the northern end of the Lake Worth lagoon system, at MacArthur State Park, Palm Beach County, Florida. Note the small rocky outcrops of the Anastasia Formation along the shore on the right. Typically, these limestone substrates harbor the Palm Beach Provinciatone endemic gastropods *Modulus pacei*, *Cerithium lindae*, and *Stramonita buchecki*. The intervening muddy sand substrates provide the habitat for the endemic lagoonal bivalves *Semele donovani* and *Mercenaria hartae* and the large predatory gastropods *Melongena (Rexmela) corona winnerae* and *Neverita delessertiana patriceae*. (Photograph by Eddie Matchett.)

an interesting fauna of hard substrate–loving endemic gastropods, including the muricid *Stramonita buchecki* (Figure 6.12C, D, K, L; also found in sea grass beds in Lake Worth); the cerithiid *Cerithium lindae* (Figure 6.12A, B); and the neritid *Nerita (Theliostyla) lindae* (shown on Figure 2.8I, J in Chapter 2; also found in the Florida Keys). Both *Stramonita buchecki* and *Cerithium lindae* have very small ranges along the southeastern Florida coast, extending only from the St. Lucie River estuary at Stuart, Martin County, southward to southern Palm Beach County.

Of special interest on the intertidal mudflats of these mangrove-lined coastal lagoons are the provinciatonal endemic bivalves *Semele donovani* (Figure 6.13B) and *Mercenaria hartae* (Figure 6.13A), both of which range only from the St. Lucie River estuary southward to southern Palm Beach County. The diminutive endemic *Mercenaria hartae* lives sympatrically with the much larger and more wide-ranging *Mercenaria campechiensis* and is morphologically similar to *Mercenaria texana*, another small venerid that is restricted to the coastal lagoons of Texas (in the Texan Molluscan Subprovince of the Carolinian Molluscan Province; see Petuch, 2013: 16, 42, 225–226). Both the Texan *Mercenaria texana* and the southeastern Floridian *Mercenaria hartae* have evolved within enclosed coastal lagoons and represent specialized evolutionary offshoots that are capable of living in the highly variable environmental conditions that are typical of these extensive estuarine systems. In deeper-water areas of the Palm Beach coastal lagoons, *Mercenaria hartae* often lives in association with beds of the Oval Seagrass (*Halophila ovalis*), an introduced Indo-Pacific marine angiosperm that ranges from Fort Pierce to the Lake Worth Lagoon. This exotic invasive sea grass was originally thought to be an endemic Floridian species and was given the name *Halophila johnsonii*.

Figure 6.12 Coastal Lagoon Mollusks of the Palm Beach Provinciatone. A, B = *Cerithium lindae* Petuch, 1987, length 12 mm. It is endemic to the coastal lagoons and rocky shorelines of Martin and Palm Beach Counties and differs from the widespread *Cerithium lutosum* (Figure 3.10F, G, Chapter 3) in that it is a narrower and much more elongated shell with a paler shell color composed of a white background with black speckles. C, D = *Stramonita buchecki* Petuch, 2013, holotype, length 28 mm. This small muricid, with its bright orange aperture, has also been found in sea grass beds within the Lake Worth Lagoon. This endemic muricid is most similar to the sympatric *S. rustica* (Figure 3.7H, Chapter 3) but differs in having an orange aperture instead of white and in having a wider shell with a broader shoulder. Both *S. buchecki* and *S. rustica* occur together with *S. floridana* in Lake Worth but can easily be separated from their congener by their smaller shell sizes. E, F = *Modulus pacei* Petuch, 1987, width 14.5 mm. This distinctive endemic species is the largest modulid found in Florida. Unlike its grass-dwelling congeners from other areas in Florida, *Modulus pacei* lives on exposed rocky outcrops and algae-covered rocks in Lake Worth. This Palm Beach Provinciatone endemic is often incorrectly referred to as *Modulus modulus* (Linnaeus, 1758) by many workers, but that species is actually a smaller and more heavily knobbed shell that is restricted to the southern Caribbean. G–I = *Neverita delessertiana patriceae* Petuch and Myers, new subspecies. Holotype, height 52 mm; dorsal, ventral, and basal views. See the Systematic Appendix at the end of the book. Compare with the nominate subspecies from quartz sand substrates in the Gulf of Mexico, *Neverita delessertiana* (Recluz, 1843) (Figure 7.16E–G, Chapter 7). J = *Melongena (Rexmela) corona winnerae* Petuch, 2004, length 147 mm. This southeastern Florida subspecies of the western Florida *Melongena corona* is found only in coastal lagoons and tidal creeks, from the St. Lucie Inlet to Boynton Beach, and is the largest melongenid found along southeastern Florida. Also known as a late Pleistocene fossil (Ft. Thompson Formation) in Palm Beach County (see Petuch and Roberts, 2007, Figure 7.6G). K, L = *Stramonita buchecki* Petuch, 2013, length 30 mm. Found on oysters near mudflats and Red Mangroves at Pine Point, northern side of Singer Island, Lake Worth Lagoon.

Some of the more biogeographically important mollusks that are endemic to the coastal lagoons of the Palm Beach Provinciatone of the Carolinian Molluscan Province are listed here by feeding type and ecological niche preferences.

1. HERBIVORES (including algal film grazers)
Gastropoda
Neritidae
Nerita (Theliostyla) lindae Petuch, 1988 (see Chapter 2)
Cerithiidae
Cerithium lindae Petuch, 1987 (Figure 6.12A, B)
Modulidae
Modulus calusa foxhalli Petuch and Myers, new subspecies. (Figure 6.13H) (Turtle Grass beds in Lake Worth)

Modulus pacei Petuch, 1987 (Figure 6.12E, F) (Often, this is incorrectly referred to as *Modulus modulus*, which is a different species that is restricted to the southern Caribbean; this large endemic species lives only on exposed rocky outcrops in Lake Worth.)
Strombidae
Eustrombus gigas (Linnaeus, 1758) form *verrilli* McGinty, 1946 (Ranging from Fort Pierce to Boynton Beach, this northern subspecies or form of *gigas* differs from typical shells by lacking spines on the body whorl, in having more numerous and upward-pointing spines on the spire, and in having a pointed posterior end on the lip; see Petuch and Sargent, 2011: 120.)

Figure 6.13 Coastal Lagoon Mollusks from the Palm Beach Provinciatone. A = *Mercenaria harlue* Petuch, 2013, holotype, width 44.3 mm. This endemic lagoonal species occurs together with the much larger *Mercenaria campechiensis* (Figure 6.2E) but differs in that it is a smaller, broader, and much flatter shell with stronger concentric lamellae; in having a characteristic smooth, shiny patch on the shell dorsum near the commissure; and in that it is a canary yellow color with purple patches along the escutcheon. This small, thin-shelled species is similar to *Mercenaria texanum* from the coastal lagoons of Texas and is found only in the lagoons and tidal creeks of Martin and Palm Beach Counties. B = *Semele donovani* McGinty, 1955, length 13 mm. This pure white Palm Beach endemic has a large, smooth, unsculptured area that covers most of the shell; it is also known as a late Pleistocene fossil (Ft. Thompson Formation) in Palm Beach County. C = *Olivella elongata* Marrat, 1871, length 10 mm. This high-spired species, which is common near Peanut Island in Lake Worth, is often incorrectly synonymized with the similar, but less elongated and less colorful, *Olivella floralia*. D = *Cyclinella tenuis* (Recluz, 1852), width 36 mm. Giant variety from the intertidal sand flats in Lake Worth. E = *Divaricella dentata* (Wood, 1815), width 30 mm. This is a Palm Beach coastal lagoon variant with bright pinkish-red umbos and early growth areas. F = *Angulus sybariticus* (Dall, 1881), length 13 mm. G = *Parvilucina costata* (d'Orbigny, 1842), width 10 mm. H = *Modulus calusa foxhalli* Petuch and Myers, new subspecies Holotype, width 12 mm. (See Systematic Appendix.). I, J = *Prunum evelynae* (Bayer, 1943), length 22 mm. This endemic Palm Beach Provinciatone marginellid is similar to the widespread western Atlantic *Prunum carneum* (Storer, 1837) but differs in that it is a larger, broader, and more inflated shell with a more heavily callused spire area. Note the large brown spot on the outer lip. K, L = *Prunum apicinum* (Menke, 1828), length 11 mm. This has a typical color form, with three brown spots on the outer edge of the lip and one brown spot near the spire; compare with the stocky, pure white Florida Keys subspecies *Prunum apicinum virgineum* (Jousseaume, 1875) (Figure 3.9G, H, Chapter 3). Collected at Pine Point, Singer Island, Lake Worth Lagoon. M = *Prunum guttatum nobilianum* (Bayer, 1943), length 25 mm. This endemic Palm Beach Provinciatone marginellid is similar to the widespread western Atlantic *Prunum guttatum* (Dillwyn, 1817) but differs in that it has a larger, heavier, broader, and more calloused shell.

2. CARNIVORES
Gastropoda
2a. GENERAL CARNIVORES/SCAVENGERS
Olivellidae
> *Olivella elongata* Marrat, 1871 (Figure 6.13C) (often misidentified as *Olivella floralia*)

Olividae
> *Americoliva bollingi* (Clench, 1943) (Figure 6.8A–D; common around Peanut Island)
> *Americoliva sayana* (Ravenel, 1834) (common around Peanut Island)

Marginellidae
> *Prunum apicinum* (Menke, 1828) (Figure 6.13K, L) (typical dark-banded form with three brown spots on the lip; compare with the pure white *Prunum apicinum virgineum* from the Florida Keys)
> *Prunum evelynae* (Bayer, 1943) (Figure 6.13I, J) (endemic to sand bottom areas, in 10- to 50-m depths, offshore of the Palm Beach lagoon system; for an illustration of the living animal, see Petuch, 2013: 44)
> *Prunum guttatum nobilianum* (Bayer, 1943) (Figure 3.13M) (endemic to sand bottom areas, in 10- to 50-m depths, offshore of the Palm Beach lagoon system)

2b. MOLLUSCIVORES
Melongenidae
> *Melongena (Rexmela) corona winnerae* Petuch, 2004 (Figure 6.12J)

2c. MOLLUSCIVORES (drilling predators)
Naticidae
> *Neverita delessertiana patriceae* Petuch and Myers, new subspecies (Figure 6.12G–I)

Muricidae-Rapaninae
> *Stramonita buchecki* Petuch, 2013 (Figure 6.12C, D, K, L)

2d. CNIDARIAN FEEDERS (suctorial on zoantherians)
Epitoniidae
> *Alexania floridana* (Pilsbry, 1945) (found under rocks at Peanut Island)

3. SUSPENSION/FILTER FEEDERS
Bivalvia
Veneridae
> *Cyclinella tenuis* (Recluz, 1852) (Figure 6.13D) (giant variant from Lake Worth)
> *Mercenaria hartae* Petuch, 2013 (Figure 6.13A) (This lagoonal endemic is smaller and flatter than *M. campechiensis* and has a broader and more oval-shaped shell with a distinctive smooth, shiny oval patch near the commisure; most specimens are a pale canary yellow color.)

Lucinidae
> *Divaricella dentata* (Wood, 1815) (Figure 6.13E) (Lake Worth variant with bright pink-red umbos and early shell areas)
> *Parvilucina costata* (d'Orbigny, 1842) (Figure 6.13G) (Lake Worth variant with reduced shell sculpture)

Semelidae
> *Semele donovani* McGinty, 1955 (Figure 6.13B)

Tellinidae
> *Angulus sybariticus* (Dall, 1881) (Figure 6.13F)
> *Macoma cerina* (C.B. Adams, 1845) (bright orange variant from Lake Worth)

The malacofauna of the Palm Beach coastal lagoons and immediate offshore areas has been found to be extremely rich, with over 750 species of mollusks having been collected, alone, near Peanut Island in Lake Worth (Carole Marshall, 2013, personal communication via email). The mangrove-lined tidal creeks and lagoons of Little Lake Worth, Munyon Island, and MacArthur State Park, in the North Palm Beach area of Palm Beach County, are under the influence of freshwater effluent and contain an impoverished but interesting malacofauna. Here, the endemic lagoonal bivalves *Mercenaria hartae* and *Semele donovani* occur on the intertidal mudflats, and the endemic lagoonal gastropods *Melongena (Rexmela) corona winnerae* and *Stramonita buchecki* can be found on, or near, large beds of the oysters *Crassostrea virginica*, *Crassostrea rhizophorae*, and *Ostreola equestris*.

Close-up of the Giant Sea Anemone, *Condylactis gigantea*, growing on a sponge and calcareous algae bioherm off Middle Torch Key, Lower Florida Keys. These large, bright purple and pink anthozoans are common faunal components of the massive sponge bioherms that house the *Cerodrillia clappi* Assemblage. (Photograph by Ron Bopp)

chapter seven

Molluscan faunas of the coral reef tract macrohabitat and gastropod reef macrohabitat (nearshore marine regime)

Introduction

The Coral Reef Tract of the Florida Keys, which extends from Soldier Key to the Dry Tortugas, is composed of an immense complex of linear reef platforms that is equivalent, structurally and biologically, only to the Great Barrier Reef of Belize and the linear reef tracts of the Bahama Banks. Throughout the Florida Keys, these shallow-water (1- to 20-m depths) coral aggregations are geographically constrained, existing only in a narrow, curving band lying between the sand-filled Hawk Channel and the depths of the Florida Straits. This unique biogenic geomorphological feature, the largest coral reef system in the continental United States, also houses one of the largest coral-associated malacofaunas in the entire tropical western Atlantic. To date, over 200 species of macrogastropods and macrobivalves are known from the Keys reef complexes.

Besides coral reefs, the South Florida Bight area harbors another type of biogenic hard substrate environment, one that is nearly equivalent in size to the Keys Reef Tract itself. Extending for almost 100 km along the Ten Thousand Islands of Collier and Monroe Counties, massive biohermal complexes of vermetid gastropods have formed linear geomorphological structures along the outermost keys. These worm shell "reefs," running from Cape Romano south to Cape Sable, have developed actual barrier systems, complete with back reef lagoons. This immense linear worm shell reef complex is unique in the Americas and provides yet another type of hard substrate macrohabitat within the Florida Keys region. In this chapter, the coral reef tract and worm shell reef macrohabitats are discussed individually.

Molluscan ecology of the coral reef tract

The Coral Reef Tract Macrohabitat, as defined here, is an amalgamation of several of the Coastal Marine Ecological Classification Standard (CMECS) coral macrohabitats. Because all of these coral aggregations (summarized in Chapter 1) shelter the same basic malacofauna, they are considered, collectively, to represent a single molluscan biotope. These can include isolated small, shallow-water patch reefs that are dominated by the hermatypic corals *Porites furcata* (Figure 7.1), *Porites porites*, and *Siderastrea radians* and the ahermatypic *Oculina diffusa*, or they can encompass the massive reef platforms that are dominated by hermatypic corals such as *Orbicella cavernosa*, *Montastrea annularis*, *Diploria clivosa*, *Diploria labyrinthiformis*, and *Acropora palmata* (Figure 7.2). The molluscan fauna of the reef systems is overwhelmingly dominated by gastropods, many of which live cryptically beneath individual coral heads or within piles of coral rubble. One of these cryptic species is the large

Figure 7.1 Close-up of a *Porites furcata* coral colony, exposed during an extreme low tide, off Missouri Key, Middle Florida Keys. Clusters of the sessile chamid bivalve *Chama congregata* Conrad, 1833, can be seen growing on the right side of the top branches of the coral colony. These types of coral bioherms are typical of the shallower-water, nearshore areas of the Coral Reef Tract Macrohabitat. (Photograph by E. J. Petuch.)

Figure 7.2 View of a large colony of the Elkhorn Coral, *Acropora palmata*, a typical shallow-water coral reef tract species. This well-developed colony is growing on the open reef platform off Key Largo, Upper Florida Keys, and shelters a school of the snapper *Lutjanus apodus*. Coralline environments such as these support the *Stephanoconus regius* Assemblage of the Coral Reef Tract Macrohabitat. (Photograph by Robert Myers.)

cone shell, *Stephanoconus regius*, which is completely restricted to reef platform environments and is here chosen to represent the entire reef molluscan faunule (the *Stephanoconus regius* Assemblage). Considering the size of the Keys reef malacofauna, there are surprisingly few endemic species present, with most of the resident mollusks simply classic Caribbean or tropical western Atlantic coral-associated taxa. Some of the better-known Florida Keys endemic coral reef species include gastropods such as the liotiids *Cyclostrema tortuganum* and *Arene vanhyningi*; the ovulid *Cyphoma alleneae* (which also may occur in Cuba); the muricids *Dermomurex pacei* (which may also occur in Cuba) and *Quoyula kalafuti*; and the amathinid *Amathina pacei*; and bivalves such as the pectinid *Caribachlamys mildredae* and the chamid *Chama inezae*. These are discussed and illustrated later in this chapter.

One of the most species-rich herbivorous gastropod groups found on the Keys reef systems is the family Fissurellidae (the Key Hole Limpets), which is represented by at least 14 species. One of the largest of these reef fissurellids is the striking green-color *Lucapina aegis* (Figure 7.3H), which is also among the rarest and most seldom-seen species of its family. These herbivorous gastropods, which graze on algae growing on dead or eroded coral heads, live sympatrically with a large fauna of other herbivores, including the liotiid *Arene cruentata* (Figure 7.3A); the trochids *Tegula lividomaculata* (Figure 7.3E) and *Tegula hotessieriana* (Figure 7.3F); the phasianellids *Eulithidium bellum* and *Eulithidium pterocladia*; and the cerithiid *Bayericerithium semiferrugineum* (Figure 7.4J). The undersurfaces of the dead coral slabs also harbor large growths of colonial urochordates (tunicates), most typically *Didemnum candidum*, *Clavelina picta*, and *Botryllus planus*. These colonial tunicates serve as the principal food source for an unusually large fauna of triviid gastropods, represented by at least 11 different species. Some of the more commonly encountered of these urochordate ectocommensal predators include *Pusula pediculus* (Figure 7.5A, B), *Niveria quadripunctata* (Figure 7.5C), *Pusula pullata* (Figure 7.5M), and *Hespererato maugeriae* (Figure 7.5E).

Of special interest on the Florida Keys Reef Tract is an unusually large fauna of tonnoidean gastropods of the families Ranellidae and Bursidae, collectively comprising over 12 species. The bursids, including both common-encountered species such as *Bursa (Colubrellina) cubaniana* (Figure 7.6J) and *Bursa (Lampasopsis) rhodostoma thomae* (Figure 7.6H) and rare species such as *Bursa (Lampasopsis) grayana* (Figure 7.6I) and *Bursa (Colubrellina) corrugata* (Figure 7.6F, G), feed almost entirely on polychaete annelid worms and sipunculid peanut worms such as *Phascolion cryptum*. These worm prey are anesthetized and digested extragastrically by the bursids' acid saliva and then ingested through their elongated proboscises. Members of the closely related family Ranellidae, although occasionally preying on other mollusks, are more specialized predators, feeding almost entirely on small reef-dwelling echinoderms such as the echinoid *Echinoneus cyclostomus*, the asteroid *Asterina folium*, and the ophiuroids *Ophiocoma echinata*, *Ophiocoma wendti*, and *Ophioderma appressum*. Also feeding on echinoids and asteroids, the small reef-dwelling Helmet Shells (Cassidae) *Cassis flammea* (Figure 7.7A, B) and *Cypraecassis testiculus* (Figure 7.7C, D) are direct competitors with the bursid Frog Shells for echinoderm prey.

Within the *Stephanoconus regius* Assemblage, the coral colonies themselves support a large combined fauna of cnidarian feeders in the gastropod families Ovulidae and Muricidae. The ornately sculptured Coral Shell muricids, all in the subfamily Magilinae, live attached directly to the coral heads and feed suctorially on the body fluids of the polyps. At least five species of Coral Shells are found on the Florida Reef Tract, with the most commonly encountered forms being *Coralliophila caribaea* (Figure 7.4A), *Coralliophila galea* (Figure 7.4C), *Coralliophila aberrans*, *Babelomurex scalariformis* (Figure 7.4E), and *Quoyula kalafuti* (Figure 7.8K, L). This last-mentioned species is the only known Atlantic representative of the tropical Pacific Ocean genus *Quoyula*, and it is presently thought to be endemic to the

Figure 7.3 Gastropods of the *Stephanoconus regius* Assemblage. A = *Arene cruentata* (Mühlfeld, 1829), width 12 mm. B = *Arene vanhyningi* Rehder, 1943, width 11 mm; endemic to the Florida Keys. C = *Calliostoma pulchrum* (C.B. Adams, 1850), height 13.2 mm. D = *Calliostoma jujubinum* (Gmelin, 1791), height 19 mm. E = *Tegula (Agathistoma) lividomaculata* (C.B. Adams, 1845), width 15.4 mm. F = *Tegula (Agathistoma) hotessieriana* (d'Orbigny, 1842), width 5 mm. G = *Calliostoma javanicum* (Gmelin, 1791), height 21 mm. H = *Lucapina aegis* (Reeve, 1850), length 28 mm. I = *Polygona carinifera* (Lamarck, 1822), length 61 mm. J = *Polygona angulata* (Röding, 1798), length 42 mm. K = *Fenimorea halidorema* (Schwengel, 1940), length 27 mm. L = *Zafrona pulchella* (Blainville, 1829), length 7 mm. M = *Pilsbryspira albocincta* (C.B. Adams, 1845), length 15 mm.

Figure 7.4 Gastropods of the *Stephanoconus regius* Assemblage. A = *Coralliophila caribbaea* Abbott, 1958, length 27 mm. B = *Bailya intricata* (Dall, 1884), length 14 mm. C = *Coralliophila galea* (Dillwyn, 1823), length 28 mm. D = *Neodrillia moseri* (Dall, 1889), length 28.5 mm. E = *Babelomurex scalariformis* (Lamarck, 1822), length 27 mm. [Often, this is incorrectly referred to as *Babelomurex mansfieldi* (McGinty, 1940), which is actually an early Pleistocene fossil from the Caloosahatchee Formation of southern Florida.] F = *Leucozonia lineata* Usticke, 1969, length 28 mm (most common on Palm Beach reefs). G = *Gemophos auritulus* (Link, 1807), length 27 mm. H = *Bailya parva* (C.B. Adams, 1850), length 14 mm. I = *Engina turbinella* (Kiener, 1835), length 9 mm. J = *Bayericerithium semmiferrugineum* (Lamarck, 1822), length 21 mm. K = *Vasum muricatum* (Born, 1778), length 62 mm. L = *Colubraria testacea* (Mörch, 1852), length 34 mm.

Figure 7.5 Cowries and Allied Cowries of the *Stephanoconus regius* Assemblage. A, B = *Pusula pediculus* (Linnaeus, 1758), length 12 mm. C = *Niveria quadripunctata* (Gray, 1827), length 4 mm. D = *Macrocypraea zebra* (Linnaeus, 1758), length 78 mm (large reef form). E = *Hespererato maugeriae* (Gray, 1832), length 3 mm. F = *Cyphoma alleneae* Cate, 1973, length 33.5 mm; lives on *Porites* corals. G, H = *Macrocypraea (Lorenzicypraea) cervus* (Linnaeus, 1771), length 101 mm. I, J = *Erosaria acicularis* (Gmelin, 1791), length 24 mm. K, L = *Luria cinerea* (Gmelin, 1791), length 24 mm. M = *Pusula pullata* (Sowerby II, 1870), length 9 mm.

Figure 7.6 Gastropods of the *Stephanoconus regius* Assemblage. A, B = *Charonia variegata* (Lamarck, 1816), length 258 mm. C = *Cymatium (Gutturnium) muricinum* (Röding, 1798), length 22 mm. D = *Cymatium (Gutturnium) nicobaricum* (Röding, 1798), length 32 mm. E = *Cymatium (Tritoniscus) labiosum* (Wood, 1828), length 21 mm. F, G = *Bursa (Colubrellina) corrugata* (Perry, 1811), length 53 mm. H = *Bursa (Lampasopsis) rhodostoma thomae* (d'Orbigny, 1842), length 26 mm. I = *Bursa (Lampasopsis) grayana* Dunker, 1862, length 25 mm. J = *Bursa cubaniana* (d'Orbigny, 1842), length 51 mm. K = *Dolicholatirus cayohuesonicus* (Sowerby III, 1879), length 13 mm. L = *Calliostoma tampaense* (Conrad, 1846), height 19 mm (also found near sea grass beds).

Figure 7.7 Gastropods of the *Stephanoconus regius* Assemblage. A, B = *Cassis flammea* (Linnaeus, 1758), length 83 mm. C, D = *Cypraecassis testiculus* (Linnaeus, 1758), length 33 mm. E = *Neodrillia cydia* (Bartsch, 1943), length 23 mm. F = *Tritonoharpa lanceolata* (Menke, 1828), length 15 mm. G, H = *Cymatium* (*Monoplex*) *pileare* (Linnaeus, 1758), length 44 mm. I, J = *Cymatium* (*Monoplex*) *aquatile* (Reeve, 1844), length 73 mm. K, L = *Cymatium* (*Septa*) *occidentale* Clench and Turner, 1947, length 31 mm.

Figure 7.8 Muricid gastropods of the *Stephanoconus regius* Assemblage. A, B = *Dermomurex pacei* Petuch, 1988, length 16 mm. C, D = *Dermomurex pauperculus* (C.B. Adams, 1856), length 24 mm. E, F = *Dermomurex (Gracilimurex) elizabethae* McGinty, 1940, length 18 mm. G, H = *Muricopsis oxytatus* (M. Smith, 1938), length 33 mm. I = *Tripterotyphis triangularis* (A. Adams, 1850), length 17 mm. J = *Trachypollia turricula* (Maltzan, 1884), length 8 mm. K, L = *Quoyula kalafuti* Petuch, 1987, length 12.9 mm.

Florida Keys Reef Tract. Although belonging to a group that normally lives on gorgonians, the large ovulid *Cyphoma alleneae* (Figure 7.5F) actually lives on *Porites* coral colonies and feeds directly on the exposed polyps, particularly those of the branching forms, such as *Porites divaricata* and *Porites furcata*. *Cyphoma alleneae* is the only western Atlantic ovulid known to occupy the ecological niche of direct grazing on hard corals. Another cnidarian-associated gastropod is the small endemic muricid *Dermomurex pacei* (Figure 7.8A, B), which is most commonly encountered within the interconnected cavities inside the holdfasts of large Sea Fans (*Gorgonia ventalina*). Within this labyrinth of eroded chambers, small rock-burrowing bivalves such as *Gregariella coralliophaga*, *Lamychaena hians*, and *Gastrochaena ovata* often occur in large aggregations, and these provide the principal prey items for the molluscivorous *Dermomurex pacei*.

The living coral reefs of the Florida Keys also house an interesting and distinctive fauna of sessile pectinid bivalves, comprising at least four species in the scallop genus *Caribachlamys*. These extremely ornate and colorful species live on the undersides of large slabs of dead coral and coral rubble, attached by their strong byssal threads. This *Caribachlamys* fauna, the richest known in the western Atlantic, is composed of such impressive species as *Caribachlamys sentis* (Figure 7.9A–C); *Caribachlamys imbricatus* (Figure 7.9F, G; Mikkelsen and Bieler, 2008, incorrectly referred to this species as *Caribachlamys pellucens*, a dubious name); *Caribachlamys mildredae* (Figure 7.9H, I; endemic to the Florida Keys and the Palm Beach coral coast); and *Caribachlamys ornatus* (Figure 7.9J, K). Along the Florida Keys Reef Tract and on the deep reefs off the Palm Beach coast, a special ecological relationship exists between these *Caribachlamys* scallops and the small amathinid gastropod *Amathina pacei* (Figure 7.9D, E). This unusual limpet-like pyramidelloidean (originally described as a *Cyclothyca* species), with its characteristic large radiating ribs, attaches itself to the auricles and bottom valves of reef-dwelling scallops, leaving behind a dark settlement spot (Figure 7.9C). This small commensal species may be feeding directly on the pectinoideans, in a typical suctorial fashion like most other pyramidelloideans, or it may actually be grazing on encrusting sponges growing on the valves of the scallops (Robert Pace, 2013, verbal communication). As far as is presently known, *Amathina pacei* is endemic to the Florida Keys and Palm Beach coast and differs from its Indo-Pacific relative *Amathina tricarinata* (Linnaeus, 1767) in having a smaller, more slender, and much more laterally compressed shell. Along Palm Beach, *Amathina pacei* has also been collected on the spines of deep-water specimens of the spiny oyster *Spondylus americanus*.

Also of special interest in the *Stephanoconus regius* Assemblage is the large fish-eating (piscivorous) cone shell *Chelyconus ermineus* (Figure 7.10L), which comes in a variety of colors, ranging from blue with black mottling to solid yellow or orange. This wide-ranging conid is found throughout the Carolinian, Caribbean, and Brazilian Molluscan Provinces and is one of the largest cone shells found on western Atlantic coral reefs. In these coralline environments, *Chelyconus ermineus* feeds on small benthonic reef fishes such as blennies, gobies, and wrasses, sneaking up on the sleeping fishes at night and harpooning them with its dart-shaped radular tooth. After injecting them with a potent venom composed of a combination of neurotoxins, hemotoxins, and proteolytic enzymes, the cone stretches its proboscis to engulf its paralyzed fish victim in a hood-like sheath. Because the Ermine Cone venom composition is specific to members of the phylum Chordata, it would also prove fatal to any human unfortunate enough to be bitten by a large specimen. For this reason, *Chelyconus ermineus* can be considered to be the only Florida Keys mollusk capable of killing a human being. This deadly piscivorous cone shell occurs together with several other harmless worm-eating cone shells, the most frequently encountered being the conids

Figure 7.9 Pectinid bivalves and ectosymbionts of the *Stephanoconus regius* Assemblage. A, B = *Caribachlamys sentis* (Reeve, 1853), length 36 mm. C = *Caribachlamys sentis* (bottom valve with *Amathina pacei* attachment spot), length 35 mm. D, E = *Amathina pacei* (Petuch, 1987), length 9.5 mm. Originally incorrectly placed in the genus *Cyclothyca*; may be feeding suctorially on scallop tissues or grazing on encrusting sponges. F, G = *Caribachlamys imbricatus* (Gmelin, 1791), length 42 mm (incorrectly referred to as *C. pellucens* by Mikkelsen and Bieler, 2008). H, I = *Caribachlamys mildredae* (Bayer, 1941), length 43 mm. J, K = *Caribachlamys ornatus* (Lamarck, 1819), length 31 mm.

Figure 7.10 Gastropods of the *Stephanoconus regius* Assemblage. A = *Morum oniscus* (Linnaeus, 1758), length 26 mm. B = *Morum purpureum* Röding, 1798, length 22 mm. C = *Vexillum variatum* (Reeve, 1845), length 19 mm. D = *Vexillum pulchellum* (Reeve, 1844), length 18 mm. E = *Vexillum dermestinum* (Lamarck, 1811), length 15 mm. F = *Scabricola nodulosa* (Gmelin, 1791), length 34 mm. G = *Scabricola pallida* (Usticke, 1959), length 32 mm. H = *Nebularia barbadensis* (Gmelin, 1791), length 33 mm. I = *Jaspidiconus mindanus* (Hwass, 1792), length 32 mm. J = *Stephanoconus regius* (Gmelin, 1791), length 50 mm. K = *Gladioconus mus* (Hwass, 1792), length 29 mm. L = *Chelyconus ermineus* (Born, 1778), length 65 mm. M = *Vexillum arestum* (Rehder, 1943), length 10 mm.

Stephanoconus regius (Figure 7.10J), *Kellyconus patae*, and *Gladioconus mus* (Figure 7.10K) and the conilithid *Jaspidiconus mindanus* (Figure 7.10I).

The vermivorous cone shells of the Florida Keys Reef Tract are in direct competition with over 17 species of sympatric worm-eating turroidean gastropods in the families Turridae, Drilliidae, and Raphitomidae. Some of the more frequently encountered of these turroidean vermivores include the zonulispirids *Pilsbryspira albocincta* (Figure 7.3M) and *Pilsbryspira jayana* and the drilliids *Fenimorea halidorema* (Figure 7.3K), *Neodrillia cydia* (Figure 7.7E), *Neodrillia moseri* (Figure 7.4D), and *Neodrillia wolfei*. The shear abundance and species richness of reef-dwelling polychaete worms has allowed the Volterra–Gauss Principle to come into play on the Florida Keys Reef Tract, permitting numerous direct competitors to coexist and flourish without detrimental ecological interactions. These small conid, conilithid, and turroidean vermivores are also in direct competition with the large sympatric turbinellid gastropod *Vasum muricatum* (Figure 7.4K), which feeds on polychaete annelids, sipunculids, and nemertean worms.

Virtually all of the bivalves of the coral reef platforms are sessile forms, living attached directly to the coral surface or by strong byssal threads, or are boring forms, living inside the coral heads in self-constructed chambers. Of the sessile species, those that are attached directly to the substrate include members of the families Gryphaeidae, Spondylidae, Plicatulidae, and Chamidae, and the byssally attached forms include members of the families Pteriidae, Pectinidae, and Malleidae. The coral rock borers include members of the families Mytilidae, Gastrochaenidae, and Trapezidae, and these coralliophagous forms are some of the major producers of bioerosion on the reef platforms. Only members of the family Limidae are free living, often swimming from one area to another by rapidly clapping their valves together, producing a weak form of jet propulsion. These spectacular pectinoidean bivalves, with their fringes of long, brightly colored mantle tentacles, often produce nests within coral rubble cavities, where they temporarily attach themselves by weak byssal threads. Of the over 200 species of macromollusks found in the *Stephanoconus regius* Assemblage, only 33 are bivalves, all of which are specialized forms that are adapted for living in and under coral slabs and coral rubble. For illustrations and discussions of the taxonomy of these bivalves, see Mikkelsen and Bieler (2008).

Biodiversity of the Stephanoconus regius *Assemblage*

The members of the coral reef-associated *Stephanoconus regius* Assemblage are listed here by feeding type and ecological niche preferences. For illustrations of the reef-dwelling gastropods that are not shown in this book, see the iconographies of Abbott (1974) and Warmke and Abbott (1962). The following resident species list is a compilation of those given in the iconographies of Abbott and Warmke and Abbott, in Petuch and Sargent (2011c), and from personal observations over a 35-year period:

1. HERBIVORES
    Gastropoda
    Lottiidae
        Patelloida pustulata (Helbling, 1779)
    Fissurellidae
        Diodora dysoni (Reeve, 1850)
        Diodora jaumei Aguayo and Farfante, 1936
        Diodora meta (von Ihering, 1927)
        Diodora sayi (Dall, 1899)

     Emarginula dentigera Heilprin, 1887
     Emarginula phrixoides Dall, 1927
     Emarginula pumila (A. Adams, 1851)
     Fissurella (Clypidella) fascicularis Lamarck, 1822
     Fissurella (Clypidella) punctata Fischer, 1857
     Hemitoma octoradiata (Gmelin, 1791)
     Hemitoma (Monfortia) emarginata (Blainville, 1825)
     Lucapina aegis (Reeve, 1850) (Figure 7.3H)
     Lucapina philippiana (Finlay, 1930)
     Lucapinella limatula (Reeve, 1850)

Trochidae
     Pseudostomatella erythrocoma (Dall, 1889)
     Tegula (Agathistoma) fasciata (Born, 1778)
     Tegula (Agathistoma) excavata (Lamarck, 1822)
     Tegula (Agathistoma) hotessieriana (d'Orbigny, 1842) (Figure 7.3F)
     Tegula (Agathistoma) lividomaculata (C.B. Adams, 1845) (Figure 7.3E)
     Synaptocochlea nigrita Rehder, 1939
     Synaptocochlea picta (d'Orbigny, 1842)

Liotiidae
     Arene cruentata (Muhlfeld, 1829) (Figure 7.3A)
     Arene vanhyningi Rehder, 1943 (Figure 7.3B)
     Cyclostrema cancellatum Marryat, 1818
     Cyclostrema tortuganum (Dall, 1927)

Turbinidae
     Lithopoma caelata (Gmelin, 1791)
     Lithopoma tuber (Linnaeus, 1758)
     Turbo (Marmarostoma) castaneu crenulatus Gmelin, 1791

Phasianellidae
     Eulithidium affinis (C.B. Adams)
     Eulithidium bellum (M. Smith, 1937)
     Eulithidium pterocladica (Robertson, 1958)

Phenacolepidae
     Phenacolepas hamillei (Fischer, 1857)

Cerithiidae
     Bayericerithium semiferrugineum (Lamarck, 1822) (Figure 7.4J)
     Cerithium guinaicum Philippi, 1849

Vanikoridae
     Vanikoro oxychone Mörch, 1877

2. CARNIVORES
Gastropoda
2a. GENERAL CARNIVORES/SCAVENGERS (multiple prey items, including hydroids and sponges)
Calliostomatidae
     Calliostoma euglyptum (A. Adams, 1854)
     Calliostoma javanicum (Gmelin, 1791) (Figure 7.3G)
     Calliostoma jujubinum (Gmelin, 1791) (Figure 7.3D)
     Calliostoma pulchrum (C.B. Adams, 1850) (Figure 7.3C)
     Calliostoma roseolum Dall, 1880
     Calliostoma scalenum Quinn, 1992

Calliostoma tampaense (Conrad, 1849) (Figure 7.6L) (also found in sea grass beds)
Calliostoma yucatecanum Dall, 1881

Cypraeidae
Erosaria acicularis (Gmelin, 1791) (Figure 7.5I, J)
Luria cinerea (Gmelin, 1791) (Figure 7.5K, L)
Macrocypraea zebra (Linnaeus, 1758) (Figure 7.5D)
Macrocypraea (Lorenzicypraea) cervus (Linnaeus, 1771) (Figure 7.5G, H)

Buccinidae
Anna florida Garcia, 2008
Bailya intricata (Dall, 1884) (Figure 7.4B)
Bailya parva (C.B. Adams, 1850) (Figure 7.4H)
Engina corinnae Crovo, 1971
Engina turbinella (Kiener, 1835) (Figure 7.4I)
Gemophos auritulus (Link, 1807) (Figure 7.4G)
Pisania pusio (Linnaeus, 1758)

Colubrariidae
Colubraria testacea (Mörch, 1852) (Figure 7.4L)

Columbellidae
Aesopus stearnsi (Tryon, 1883)
Columbella mercatoria (Linnaeus, 1758)
Columbellopsis nycteus (Duclos, 1846)
Costoanachis sparsa (Reeve, 1859) (also found in sea grass beds)
Costoanachis translirata (Ravenel, 1861)
Nassarina monilifera (Sowerby II, 1844)
Nassarina sparsipunctata (Rehder, 1943)
Nitidella laevigata (Linnaeus, 1758)
Nitidella nitida (Lamarck, 1822)
Suturoglypta hotessieriana (d'Orbigny, 1842)
Suturoglypta iontha (Ravenel, 1861)
Zafrona idalina (Duclos, 1840)
Zafrona pulchella (Blainville, 1829) (Figure 7.3L)

Harpidae
Morum oniscus (Linnaeus, 1758) (Figure 7.10A)
Morum purpureum Röding, 1798 (Figure 7.10B)

Marginellidae
Hyalina pallida (Linnaeus, 1758)
Volvarina lactea (Kiener, 1841)

2b. MOLLUSCIVORES (feeding on both bivalves and gastropods, including drillers)
Muricidae-Muricinae, Muricopsinae, Typhinae, Ergalitaxinae (the coralliopha-gous subfamily Magilinae is listed separately)
Aspella anceps (Lamarck, 1822)
Aspella castor Radwin and D'Attilio, 1976
Caribiella alveata (Kiener, 1842)
Dermomurex pacei Petuch, 1988 (Figure 7.8A, B)
Dermomurex pauperculus (C.B. Adams, 1856) (Figure 7.8C, D)
Dermomurex (Gracilimurex) elizabethae McGinty, 1940 (Figure 7.8E, F)
Favartia minirosea (Abbott, 1954)
Muricopsis oxytatus (M. Smith, 1938) (Figure 7.8G, H)
Pterotyphis pinnatus (Broderip, 1833)

Risomurex caribbaeus (Bartsch and Rehder, 1939)
Risomurex deformis (Reeve, 1846)
Risomurex muricoides (C.B. Adams, 1845)
Risomurex roseus (Reeve, 1846)
Trachypollia turricula (Maltzan, 1884) (Figure 7.8J)
Tripterotyphis triangularis (A. Adams, 1856) (Figure 7.8I)

Fasciolariidae
Dolicholatirus cayohuesonicus (Sowerby III, 1879) (Figure 7.6K)
Leucozonia lineata Usticke, 1969 (Figure 7.4F)
Leucozonia ocellata (Gmelin, 1791)
Polygona angulata (Röding, 1798) (Figure 7.3J)
Polygona carinifera (Lamarck, 1822) (Figure 7.3I)
Polygona infundibula (Gmelin, 1791)
Polygona nemata (Woodring, 1928)

Costellariidae
Pusia puella (Reeve, 1845)
Thala foveata (Sowerby II, 1874)
Vexillum arestum (Rehder, 1943) (Figure 7.10M)
Vexillum dermestinum (Lamarck, 1811) (Figure 7.10E)
Vexillum epiphanea Rehder, 1943
Vexillum histrio (Reeve, 1844)
Vexillum moisei McGinty, 1955
Vexillum moniliferum (C.B. Adams, 1850) (often referred to as *V. albocinctum*, which
    is a nomen dubium)
Vexillum pulchellum (Reeve, 1844) (Figure 7.10D)
Vexillum sykesi (Melvill, 1925)
Vexillum variatum (Reeve, 1845) (Figure 7.10C)

2c. CNIDARIAN FEEDERS (including living corals and zoantherians)
Epitoniidae
Alexania floridana (Pilsbry, 1945)
Asperiscala apiculata (Dall, 1889)
Asperiscala candeana (d'Orbigny, 1842)
Asperiscala denticulata (Sowerby I, 1844)
Asperiscala multistriata (Say, 1826)
Asperiscala novangliae (Couthouy, 1838)
Cycloscala echinaticosta (d'Orbigny, 1842)
Dentiscala crenata (Linnaeus, 1758)
Dentiscala hotessieriana (d'Orbigny, 1842)
Depressiscala nautlae (Mörch, 1874)
Epitonium albidum (d'Orbigny, 1842)
Epitonium angulatum (Say, 1830)
Epitonium championi Clench and Turner, 1952
Epitonium foliaceicostum (d'Orbigny, 1842)
Epitonium krebsii (Mörch, 1874)
Epitonium humphreysi (Kiener, 1838)
Epitonium occidentale (Nyst, 1871)
Epitonium tollini Bartsch, 1938
Epitonium unifasciatum (Sowerby I, 1844)
Gyroscala lamellosa (Lamarck, 1822)

Nodiscala pumilio (Mörch, 1874)
Pictoscala rupicola (Kurtz, 1860)
Spiniscala echinaticosta (d'Orbigny, 1842)
Architectonicidae
Heliacus bisulcatus (d'Orbigny, 1842)
Heliacus cylindrus (Gmelin, 1791)
Ovulidae
Cyphoma alleneae Cate, 1973 (Figure 7.5F)
Muricidae-Magilinae
Babelomurex scalariformis (Lamarck, 1822) (Figure 7.4E)
Coralliophila aberrans (C.B. Adams, 1850)
Coralliophila caribbaea Abbott, 1958 (Figure 7.4A)
Coralliophila galea (Dillwyn, 1823) (Figure 7.4C)
Quoyula kalafuti Petuch, 1987 (Figure 7.8K, L)
2d. UROCHORDATE FEEDERS (including colonial tunicates)
Lamellariidae
Lamellaria leucosphaera Schwengel, 1942
Lamellaria perspicua (Linnaeus, 1758)
Triviidae
Cleotrivia candidula (Gaskoin, 1836)
Dolichupis leei Fehse and Grego, 2002
Dolichupis leucosphaera Schilder, 1931
Hespererato maugeriae (Gray, 1832) (Figure 7.5E)
Niveria antillarum (Schilder, 1922)
Niveria maltbiana (Schwengel and McGinty, 1942)
Niveria nix (Schilder, 1922) (Figure 8.3, see Chapter 8)
Niveria quadripunctata (Gray, 1827) (Figure 7.5C)
Pusula pediculus (Linnaeus, 1758) (Figure 7.5A, B)
Pusula pullata (Sowerby II, 1870) (Figure 7.5M)
Pusula suffusa (Gray, 1832)
2e. ECHINODERM FEEDERS (including echinoids, ophiuroids, and asteroids)
Ranellidae
Charonia variegata (Lamarck, 1816) (Figure 7.6A, B)
Cymatium (Cymatriton) nicobaricum (Röding, 1798) (Figure 7.6D) (occasionally feeds on small gastropods)
Cymatium (Gutturium) muricinum (Röding, 1798) (Figure 7.6C)
Cymatium (Monoplex) aquatile (Reeve, 1844) (Figure 7.7I, J)
Cymatium (Monoplex) parthenopeum (von Salis, 1793)
Cymatium (Monoplex) pileare (Linnaeus, 1758) (Figure 7.7G, H)
Cymatium (Septa) occidentale Clench and Turner, 1947 (Figure 7.7K, L)
Cymatium (Septa) vespaceum (Lamarck, 1822)
Cymatium (Tritoniscus) labiosum (Wood, 1828) (Figure 7.6E)
Cassidae (reef-dwelling species)
Cassis flammea (Linnaeus, 1758) (Figure 7.7A, B)
Cypraecassis testiculus (Linnaeus, 1758) (Figure 7.7C, D)
2f. SUCTORIAL FEEDERS
Cancellariidae (problematical; elasmobranchs and invertebrates?)
Bivetopsia rugosum (Lamarck, 1822)
Tritonoharpa lanceolata (Menke, 1828) (Figure 7.7F)

2g. PISCIVORES (feeding on blennies and gobies)
Conidae-Coninae
  Chelyconus ermineus (Born, 1778) (Figure 7.10L)
2h. VERMIVORES (including polychaetes, sipunculids, and nemerteans)
Bursidae
  Bursa (Colubrellina) corrugata (Perry, 1811) (Figure 7.6F, G)
  Bursa (Colubrellina) cubaniana (d'Orbigny, 1842) (Figure 7.6J)
  Bursa (Lampasopsis) grayana Dunker, 1862 (Figure 7.6I)
  Bursa (Lampasopsis) rhodostoma thomae (d'Orbigny, 1842) (Figure 7.6H)
Mitridae
  Nebularia barbadensis (Gmelin, 1791) (Figure 7.10H)
  Scabricola nodulosa (Gmelin, 1791) (Figure 7.10F)
  Scabricola pallida (Usticke, 1959) (Figure 7.10G)
Turbinellidae
  Vasum muricatum (Born, 1778) (Figure 7.4K)
Conidae-Puncticulinae
  Gladioconus mus (Hwass, 1792) (Figure 7.10K)
  Kellyconus patae (Abbott, 1971)
  Stephanoconus regius (Gmelin, 1791) (Figure 7.10J)
Conilithidae
  Jaspidiconus mindanus (Hwass, 1792) (Figure 7.10I)
Crassispiridae
  Crassispira rhythmica Melvill, 1927
Strictispiridae
  Strictispira solida (C.B. Adams, 1850) (= *fuscesens* "Reeve" of authors)
Zonulispiridae
  Pilsbryspira albocincta (C.B. Adams, 1845) (Figure 7.3M)
  Pilsbryspira jayana (C.B. Adams, 1850)
  Pilsbryspira monilis (Bartsch and Rehder, 1939)
  Pilsbryspira nodata (C.B. Adams, 1850) (*albomaculata* is a synonym)
  Zonulispira crocata (Reeve, 1845) (= *sanibelensis* Bartsch and Rehder, 1939)
Drilliidae
  Cerodrillia perryae Bartsch and Rehder, 1939
  Fenimorea halidorema Schwengel, 1940 (Figure 7.3K)
  Neodrillia cydia (Bartsch, 1943) (Figure 7.7E)
  Neodrillia moseri (Dall, 1889) (Figure 7.4D)
  Neodrillia wolfei (Tippett, 1995)
Cochlespiridae
  Pyrgospira ostrearum (Stearns, 1872)
  Pyrgospira tampaensis (Bartsch and Rehder, 1939)
Raphitomidae
  Cryoturris fargoi McGinty, 1955
  Daphnella lymnaeiformis (Kiener, 1840)
  Ithycythara auberiana (d'Orbigny, 1847)
  Ithycythara lanceolata (C.B. Adams, 1850)
  Ithycythara parkeri Abbott, 1958
  Mangelia astricta Reeve, 1846
  Pyrgocythara candidissima (C.B. Adams, 1845)

2i. PROBLEMATICAL (possibly feeding suctorially on pectinoideans or eating sponges)
  Amathina pacei (Petuch, 1987) (Figure 7.9D, E)
3. SUSPENSION/FILTER FEEDERS
Gastropoda
Vermetidae
  Dendropoma corrodens (d'Orbigny, 1841) (= "*annulatus*")
  Dendropoma irregulare (d'Orbigny, 1841)
  Petaloconchus erectus (Dall, 1888)
  Petaloconchus mcgintyi (Olsson and Harbison, 1953)
  Petaloconchus varians (d'Orbigny, 1841)
  Serpulorbis decussatus (Gmelin, 1791)
Capulidae
  Capulus (Krebsia) intortus (Lamarck, 1822)
Hipponicidae
  Hipponix antiquatus (Linnaeus, 1767)
  Hipponix subrufus (Lamarck, 1822)
Calyptraeidae
  Calyptraea centrulis (Conrad, 1841)
  Cheilea equestris (Linnaeus, 1758)
  Crucibulum auricula (Gmelin, 1791)
  Crucibulum striatum Say, 1824
Bivalvia
Nuculidae
  Nucula calcicola D. Moore, 1977 (only known reef-dwelling nuculid)
Mytilidae
  Gregariella coralliophaga (Gmelin, 1791)
  Lithophaga antillarum (d'Orbigny, 1853)
  Lithophaga aristata (Dillwyn, 1817)
  Lithophaga bisulcata (d'Orbigny, 1853)
  Lithophaga nigra (d'Orbigny, 1853)
  Modiolus squamosus Beauperthuy, 1967
Pteriidae
  Pinctada imbricata Röding, 1798
  Pteria colymbus (Röding, 1798)
Trapezidae
  Coralliophaga coralliophaga (Gmelin, 1791)
Malleidae
  Malleus candeanus (d'Orbigny, 1853) (Figure 7.11H)
Gryphaeidae
  Hyotissa mcgintyi (Harry, 1985) (Figure 7.11G)
Limidae
  Ctenoides mitis (Lamarck, 1807)
  Ctenoides scabra (Born, 1778)
  Lima caribaea d'Orbigny, 1853 (Figure 7.11K)
  Limaria pellucida (C.B. Adams, 1846) (Figure 7.11D)
Pectinidae
  Caribachlamys imbricatus (Gmelin, 1791) (Figure 7.9F, G) (Note: Mikkelsen and
    Bieler, 2008, incorrectly identified this species as *Caribachlamys pellucens*.)
  Caribachlamys mildredae (Bayer, 1941) (Figure 7.9H, I)

Figure 7.11 Bivalves of the *Stephanoconus regius* Assemblage. A = *Chama macerophylla* Gmelin, 1791, length 69 mm. B = *Chama florida* Lamarck, 1819, length 32 mm. C = *Chama inezae* (Bayer, 1943), length 74 mm. D = *Limaria pellucida* (C.B. Adams, 1846), length 23 mm. E = *Plicatula gibbosa* Lamarck, 1801, length 29 mm. F = *Spondylus tenuis* Schriebers, 1793, length 62 mm (= *S. ictericus*). G = *Hyotissa mcgintyi* (Harry, 1985), length 64 mm. H = *Malleus candeanus* (d'Orbigny, 1853), length 39 mm. I = *Lima caribaea* d'Orbigny, 1853, length 42 mm.

Caribachlamys ornatus (Lamarck, 1819) (Figure 7.9J, K)
Caribachlamys sentis (Reeve, 1853) (Figure 7.9A–C)
Spondylidae
Spondylus americanus Hermann, 1781
Spondylus tenuis Schriebers, 1793 (Figure 7.11F) (*ictericus* Reeve, 1856, is a synonym)
Plicatulidae
Plicatula gibbosa Lamarck, 1801 (Figure 7.11E)
Chamidae
Chama congregata Conrad, 1833 (Figure 7.1)
Chama florida Lamarck, 1819 (Figure 7.11B)
Chama inezae (Bayer, 1843) (Figure 7.11C)
Chama macerophylla Gmelin, 1791 (Figure 7.11A, B)
Chama radians Lamarck, 1819
Chama sarda Reeve, 1847
Chama sinuosa Broderip, 1835
Gastrochaenidae
Gastrochaena ovata (Sowerby I, 1834)
Lamychaena hians (Gmelin, 1791)
Spengleria rostrata (Spengler, 1783)

A review of the ecological niche preferences of the members of the *Stephanoconus regius* Assemblage shows that the carnivorous gastropods, including specialized types such as molluscivores, echinoderm feeders, urochordate feeders, coral feeders, and fish eaters, overwhelmingly dominate the coral reef platform ecosystem.

Molluscan ecology of the vermetid bioherms

The immense chain of worm shell bioherms ("reefs") found along the Outer Ten Thousand Islands of the South Florida Bight (discussed in Chapter 1 and previously in this chapter) is unique in the western Atlantic region. Extending from Cape Romano in the north to Cape Sable in the south, these reefs are composed entirely of a mono-culture of the worm gastropod *Vermetus* (*Thylaeodus*) *nigricans*, a sessile species that produces massive intertwined pseudocolonies. Interestingly, these unusual large-scale biogenic geomorphological features form complex structures that mimic actual coral reefs. On some of the more well-developed and intricate worm shell reef systems, such as those off Demijohn Key (Figure 7.12), the pseudocolonies of vermetids form promi-nent wide reef platforms that are exposed at low tide. These platforms are often edged by barrier-type reefs (linear bioherms) that produce wide, shallow lagoons and numerous deep tide pools (Figure 7.13). These barrier reef and wide reef platforms are especially well developed on the islands near the Chokoloskee and Sandfly Passes, particularly off Jewell, Panther, Pavilion, Round, Tiger, Indian, Turtle, and Rabbit Keys. The vermetid barrier bioherms typically grow along the windward ocean sides of the outermost keys of the entire archipelago and are exposed to nearly continuous low-level wave action. On some islands, such as Jewell Key, the beaches are composed almost entirely of disin-tegrated vermetid pseudocolonies, the result of damage produced during strong storms and heavy wave action.

Like their coral reef counterparts, these vermetid structures exhibit a deterministic developmental pattern, growing outward into the shallow wave zone. This continuous seaward growth provides a substrate for the colonization of mangroves, with the outer keys having first developed as extensive worm shell reefs and later having been overgrown by Red and Black Mangrove forests. On many of the outer keys of the Ten Thousand Islands, such as Rabbit Key, the vermetid reefs have been overgrown by immense aggregations of the oysters *Crassostrea virginica* and *Ostreola equestris*. These massive accumulations of ostreid bivalves, growing on cores of dead worm shell bioherms, form extensive banks that extend around onto the sheltered lee sides of the islands. These oyster banks also form large biohermal structures that serve as substrates for mangrove colonization. As described by Shier (1969), this pattern of ecological succession, from the pioneer vermetid reefs to the intermediate oyster banks to the climax mangrove forests, appears to be the primary land-building mechanism along the seaward edge of the Reticulated Coastal Swamps and the Outer Ten Thousand Islands.

The tidal pools on the Ten Thousand Islands vermetid reefs house an interesting, but impoverished, molluscan fauna, containing several special localized ecomorphs of typical and widespread southern Floridian species. Primary among these is a bright red color form of the Tulip Shell, *Fasciolaria tulipa* (Figure 7.14C), which is the primary molluscan predator of the abundant *Batillaria minima* snails and small *Ostreola equestris* oysters. The suspension-feeding *Vermetus (Thylaeodus) nigricans* worm shells (Figure 7.14A), the main reef builders and namesake of the entire molluscan faunule (the *Vermetus nigricans* Assemblage), are also one of the principal prey items of the molluscivorous buccinid gastropod *Gemophos tinctus pacei* (Figure 7.14D, E). This small whelk, a subspecies of the wide-ranging western Atlantic *Gemophos tinctus*, is endemic to the Ten Thousand Islands and occurs abundantly on some of the worm reef platforms and associated oyster ridges. Also occurring with *Gemophos tinctus pacei* on the vermetid reef complexes is the turritellid gastropod *Vermicularia fargoi owensi* (Figure 7.14F, G; see Systematic Appendix), a subspecies of the common western Florida *Vermicularia fargoi*. This small turritellid worm shell is also endemic to the Ten Thousand Islands and lives on open expanses of sand within the shallow back-reef lagoons of the vermetid barrier bioherms.

Biogeographically, the molluscan assemblages of the Ten Thousand Islands area belong to the Suwannean Molluscan Subprovince of the Carolinian Molluscan Province and share more taxa in common with the fauna of that cooler-water biotic subdivision than they do with the fauna of the tropical Floridian Molluscan Subprovince of the Carolinian Molluscan Province (the Florida Keys area; see Petuch, 2013). A classic example of this western Floridian Suwannean influence is demonstrated by the presence of the Crown Conch, *Melongena (Rexmela) corona corona* (Figure 7.14I), which ranges only as far south as the Ten Thousand Islands and Cape Sable. Here, this characteristic Suwannean gastropod, represented by a small, darkly colored local population, lives within the back-reef lagoons of the vermetid reefs in association with the endemic turritellids and buccinids. From Cape Sable southward into Florida Bay and the Florida Keys, *Melongena (Rexmela) corona corona* is replaced by the Keys endemic *Melongena (Rexmela) bicolor*, and the two species do not appear to cooccur at any known locality. Other typical Suwannean gastropods that live in the sandy lagoons and tide pools along with the Crown Conch are the voracious molluscivorous busyconids *Fulguropsis pyruloides* (Figure 7.14J) and *Sinistrofulgur sinistrum*; the modulid *Modulus floridanus* (Figure 7.14K); the fasciolariid *Cinctura hunteria*; the strombid *Strombus alatus*; the conid *Gradiconus anabathrum* (Figure 7.14H); the conilithid *Jaspidiconus stearnsi* (Figure 7.14L); and the terebrids *Strioterebrum dislocatum* (Figure 7.14B) and *Strioterebrum vinosum* (Figure 7.14M).

Figure 7.12 View of a vermetid gastropod bioherm ("reef") exposed at low tide on Demijohn Key, Ten Thousand Islands, Collier County, Florida. In the distance are two larger islands of the archipelago, Lumber Key and Rabbit Key, both of which contain large vermetid worm shell reefs along their seaward sides. This type of environment supports the *Vermetus nigricans* Assemblage of the Gastropod Reef Macrohabitat. (Photograph by E. J. Petuch.)

Figure 7.13 Close-up of a red color form of the Tulip Shell *Fasciolaria tulipa* (Linnaeus, 1758) and valves of the oyster *Ostreola equestris* (Say, 1834) in a tide pool on the vermetid reef off Demijohn Key, Ten Thousand Islands, Collier County, Florida. Large pools like this are best developed on the main platform of the worm shell reefs, behind the topographically higher barrier "reef" that grows along the seaward side of the biohermal complex. (Photograph by E. J. Petuch.)

Biodiversity of the Vermetus nigricans *Assemblage*

Both the vermetid reefs and the adjacent sandy lagoons house a species-poor macro-mollusk fauna composed of only 71 taxa. Even with the inclusion of the two endemic subspecies (*Gemophos tinctus pacei* and *Vermicularia fargoi owensi*), the *Vermetus nigricans* Assemblage still resembles a highly impoverished version of the much richer coastal lagoon faunas found along western Florida (the Suwannean Molluscan Subprovince of the Carolinian Molluscan Province). The extreme oceanographic conditions of the Ten Thousand Islands area, including seasonally fluctuating salinities and low water tempera-tures during the winter months, have prevented the establishment of a richer molluscan fauna. For the most part, the malacofauna is dominated by oysters, and their accumulated shells form most of the hardbottom substrates. Along the seaward-most keys of the Ten Thousand Islands (the Outer Ten Thousand Islands), large accumulations of vermetid reef fragments, mixed with quartz sand, make up most of the beaches, and the massive worm shell reefs of the intertidal zone offer the only extensive hard substrate areas. The seaward beaches of the outer keys are often covered with large accumulations of driftwood and weathered dead Black Mangrove trees, and these offer a substrate for a small dome com-posed of the periwinkles *Littoraria angulifera* and *Littoraria nebulosa* (and color form *tessel-lata*) (Figures 7.15 and 7.16) and the ellobiids *Melampus bidentatus* and *Ellobium* (*Auriculoides*) *dominicense* (Figure 7.16H, I).

The following species list is a compilation of both the worm reef residents and the spe-cies found on the adjacent sandy seafloors and oyster banks, and this is the first species survey ever undertaken for the Ten Thousand Islands (during 2013 and 2014). The mem-bers of the *Vermetus nigricans* Assemblage, and those of the surrounding sandy seafloors, are listed here by feeding type and ecological niche preference: **R** = living on the worm shell reef; **S** = living on sandy areas adjacent to the reef complexes.

1. HERBIVORES (including algal film grazers)
Gastropoda
Fissurellidae
   Diodora listeri (d"Orbigny, 1842) **R**
   Lucapinella suffusa (Reeve, 1850) **R**

Figure 7.14 Gastropods of the *Vermetus nigricans* Assemblage. A = *Vermetus* (*Thylaeodus*) *nigricans* (Dall, 1884), aggregation length 210 mm. B = *Strioterebrum dislocatum* (Say, 1822), length 39 mm. C = *Fasciolaria tulipa* (Linnaeus, 1758), red color form, length 66 mm. D, E = *Gemophos tinctus pacei* Petuch and Sargent, 2011, holotype, length 25.4 mm. F, G = *Vermicularia fargoi owensi* Petuch and Myers, new subspecies; holotype, length 23 mm (see Systematic Appendix). H = *Gradiconus anabathrum* (Crosse, 1865), length 43 mm. I = *Melongena* (*Rexmela*) *corona corona* (Gmelin, 1791), length 72 mm. J = *Fulguropsis pyruloides* (Say, 1822), length 68 mm. K = *Modulus floridanus* (Conrad, 1869), height 12 mm. This endemic western Florida species is often incorrectly referred to by many workers as *Modulus modulus* (Linnaeus, 1758). L = *Jaspidiconus stearnsi* (Conrad, 1869), length 20 mm. M = *Strioterebrum vinosum* (Dall, 1889), length 15 mm. With the exception of the wide-ranging *Vermetus nigricans*, *Strioterebrum dislocatum*, and *Fasciolaria tulipa*, all of these species are endemic faunal components of the Suwannean Molluscan Subprovince of the Carolinian Molluscan Province, and the Ten Thousand Islands area represents the farthest southeastern limit of their biogeographical ranges. The boundary between the Suwannean and Floridian Molluscan Subprovinces appears to be at Cape Sable, Monroe County. Here, the typical quartz sand substrates and cooler winter water condi-tions of the Suwannean Molluscan Subprovince give way to the carbonate sediments and warmer-water conditions of the Floridian Molluscan Subprovince.

Figure 7.15 Close-up of an aggregation of the littorinid *Littoraria nebulosa* (Lamarck, 1822), and its dark color form *tessellata* Philippi, 1847, crawling on a piece of driftwood along the seaward side of Rabbit Key, Ten Thousand Islands. Although relatively uncommon in southern Florida, this large periwinkle is unusually abundant on the dense driftwood accumulations found along the outer keys of the Ten Thousand Islands. The endemic Ten Thousand Islands Raccoon, *Procyon lotor marinus*, has been observed to feed on these littorinids. (Photograph by Robert Owens.)

Littorinidae

Littoraria angulifera (Lamarck, 1822) (on mangroves and higher areas of the drift-wood accumulations)

Littoraria nebulosa (Lamarck, 1822) (and color form *tessellata* Philippi, 1847; on lower areas in the splash zone of the driftwood accumulations) (Figures 7.15 and 7.16B, C)

Batillariidae

Batillaria minima (Gmelin, 1791) **R**

Cerithiidae

Cerithium lutosum Menke, 1828 **R**

Modulidae

Modulus floridanus (Conrad, 1869) (Figure 7.14K) (often incorrectly referred to as *Modulus modulus*, which is actually a southern Caribbean species; found on *Halodule* Shoal Grass) **S**

Strombidae

Strombus alatus Gmelin, 1791 **S**

Ellobiidae

Ellobium (Auricoloides) dominicense (Ferussac, 1821) (Figure 7.16H, I)

Melampus bidentatus (Say, 1822) (Both species live on driftwood accumulations, plant detritus, and Black Mangrove roots found within the interiors of the islands.)

2. CARNIVORES

Gastropoda

2a. MOLLUSCIVORES (including drilling muricids and naticids)

Naticidae

Neverita delessertiana (Recluz, 1843) (Figure 7.16E–G) **S**

Muricidae-Muricinae

Chicoreus dilectus (A. Adams, 1855) **R**

Phyllonotus pomum (Gmelin, 1791) **R**

Fasciolariidae

Cinctura hunteria (Perry, 1811) **R**

Fasciolaria tulipa (Linnaeus, 1758) (Figure 7.14C) **R**

Busyconidae

Fulguropsis pyruloides (Say, 1822) (Figure 7.14J) **S**

Sinistrofulgur sinistrum (Hollister, 1958) **S**

Melongenidae

Melongena (Rexmela) corona corona (Gmelin, 1791) (Figure 7.14I) **R**

Buccinidae

Gemophos tinctus pacei Petuch and Sargent, 2011 (Figure 7.14D, E) **R**

2b. GENERAL CARNIVORES/SCAVENGERS

Nassariidae

Phrontis vibex (Say, 1822) **S**

Columbellidae

Costoanachis semiplicata (Stearns, 1873) (Figure 7.16J, K) **S** (lives **in algae** near the worm reefs)

Olividae

Americoliva sayana (Ravenel, 1834) **S**

Marginellidae

Prunum apicinum (Menke, 1828) **S**

Bullidae

Bulla occidentalis A. Adams, 1850 **S**

2c. SPECIALIZED CARNIVORES (suctorial feeders on elasmobranchs?)

Cancellariidae

Cancellaria reticulata (Linnaeus, 1767) **S**

2d. VERMIVORES (polychaete worms)

Conidae-Puncticulinae

Gradiconus anabathrum (Crosse, 1865) (Figure 7.14H) **S**

Conilithidae-Conilithinae

Jaspidiconus stearnsi (Conrad, 1869) (Figure 7.14L) **S**

Terebridae

Strioterebrum dislocatum (Say, 1822) (Figure 7.14B) **S**

Strioterebrum vinosum (Dall, 1889) (Figure 7.14M) **S**

2e. ECHINODERM FEEDERS (holothurians)

Ficus communis Röding, 1798 (Figure 7.16D) **S**

3. SUSPENSION/FILTER FEEDERS (including detritivores)

Gastropoda

Vermetidae

Vermetus (Thylaeodus) nigricans (Dall, 1884) (Figure 7.14A) **R**

Turritellidae

Vermicularia fargoi owensi Petuch and Myers, new subspecies (Figure 7.14F, G) (see Systematic Appendix) **S** (free-living on open muddy sand bottoms near worm reefs and bryozoan colonies)

Vermicularia knorri (Deshayes, 1843) **S** (growing in sponges and bryozoan colonies)

Calyptraeidae

Crepidula fornicata (Linnaeus, 1758) **R**

Crepidula maculosa Conrad, 1846 **R**

Ianacus plana (Say, 1822) **R**

Bivalvia

Arcidae

Anadara floridana 1869) **S**

Anadara transversa (Say, 1822) **S**

Noetiidae

Arcopsis adamsi (Dall, 1886) **R**

Noetia ponderosa (Say, 1822) **S**

Figure 7.16 Mollusks of the *Vermetus nigricans* Assemblage. A= *Dinocardium vanhyningi* Clench and Smith, 1944, length 61 mm. Extremely common in the deeper channels between the Outer Ten Thousand Islands; confined to the Gulf of Mexico. B, C= *Littoraria nebulosa* (Lamarck, 1822), yellow color form, length 11 mm. D= *Ficus communis* Röding, 1798, length 75 mm. Prefers quartz sand substrates and is common along western Florida; not found in the carbonate sediment substrate areas of the Florida Keys. This species is frequently encountered around Jewell Key, where this specimen was collected. E, F, G= *Neverita delessertiana* (Recluz, 1843), dorsal, ventral, and basal views, showing the operculum within the aperture; width 43 mm. Confined to intertidal and shallow water quartz sand areas throughout the Gulf of Mexico, from Texas to the Ten Thousand Islands; not found on the carbonate substrates of the Florida Keys and coral coast. Compare with *Neverita delessertiana patriceae* (Figure 6.12G, H, I; Chapter 6), the geographically isolated, high-spired subspecies from the Palm Beach Provinciatone. H, I= *Ellobium (Auriculoides) dominicense* (Ferussac, 1821), length 13 mm. J, K = *Costoanachis semiplicata* (Stearns, 1873), length 12 mm.

Mytilidae
> *Arcuatula papyria* (Conrad, 1846) **S (in algae)**
> *Brachidontes exustus* (Linnaeus, 1758) **R**
> *Modiolus americanus* (Leach, 1815) **R**

Pinnidae
> *Atrina rigida* (Lightfoot, 1786) **S**
> *Atrina serrata* (Sowerby I, 1825) **S**

Ostreidae
> *Crassostrea rhizophorae* (Guilding, 1828) **R**
> *Crassostrea virginica* (Gmelin, 1791) **R**
> *Ostreola equestris* (Say, 1834) (Figure 7.13) **R**
>
> These three oysters are among the principal prey items of the endemic Ten
>   Thousand Islands Raccoon, *Procyon lotor marinus.*

Pectinidae
> *Argopecten irradians taylorae* Petuch, 1987 **S** (Bright orange color forms are
>   common.)

Anomiidae
> *Anomia simplex* d'Orbigny, 1853 **R**

Carditidae
> *Carditamera floridana* Conrad, 1838 **S**

Lucinidae
> *Ctena orbiculata* (Montagu, 1808) **S**
> *Phacoides pectinata* (Gmelin, 1791) **S**
> *Stewartia floridana* (Conrad, 1833) **S**

Cardiidae
> *Dallocardia muricata* (Linnaeus, 1758) **S**
> *Dinocardium vanhyningi* Clench and Smith, 1944 (Figure 7.16A) **S**
> *Laevicardium mortoni* (Conrad, 1830) **S**
> *Laevicardium serratum* (Linnaeus, 1758) **S**
> *Trachycardium egmontianum* (Shuttleworth, 1856) **S**

Veneridae
> *Anomalocardia cuneimeris* (Conrad, 1846) S
> *Chione elevata* Say, 1822 **S**
> *Macrocallista nimbosa* (Lightfoot, 1786) **S**
> *Mercenaria campechiensis* (Gmelin, 1791) **S**

Tellinidae
> *Eurytellina lineata* (Turton, 1819) **S**
> *Macoma tenta* (Say, 1834) **S**

Psammobiidae
> *Heterodonax bimaculatus* (Linnaeus, 1758) **S**

Semelidae
> *Semele proficua* (Pulteney, 1799) **S**
> *Semele purpurascens* (Gmelin, 1791) **S**

Solecurtidae
> *Tagelus divisus* (Spengler, 1794) **S**

Mactridae
> *Mactrotoma fragilis* (Gmelin, 1791) **S**

Pholadidae
> *Cyrtopleura costata* (Linnaeus, 1758) **S**

Periplomatidae

Periploma margaritaceum (Lamarck, 1801) **S**

A review of the ecological niche preferences of the members of the *Vermetus nigricans* Assemblage (the actual reef-dwelling taxa are designated by **R**) shows that the suspension/filter-feeding species dominate the ecosystem. The ancillary taxa from the surrounding sand bottom areas (designated by **S**) follow a similar pattern, with the suspension/filter-feeding species again dominating the ecosystem. The Ten Thousand Islands of extreme southwestern Florida are now known to house a highly impoverished molluscan fauna comprising only 71 species (as of this survey, for the area of the Chokoloskee and Sandfly Passes, Chokoloskee Bay, and Demijohn, Jewell, and Rabbit Keys). The nutrient-rich freshwater effluent emptying out of the Big Cypress Swamp and Everglades, through the back bays of the Reticulated Coastal Swamps, has created the ideal oceanographic conditions for the growth of extensive vermetid worm shell reefs. The spatial extent of these gastropod bioherms is intimately tied to the high phytoplankton productivity produced by the fluvial and paludal input in the area between Capes Romano and Sable, along the northern side of the South Florida Bight. The deep channels between the Outer Ten Thousand Islands and the vermetid bioherms, such as those at the mouths of Chokoloskee, Fakahatchee, and Sandfly Passes, house unusual soft-substrate communities dominated by dense growths of the ectoproct bryozoan *Schizoporella*. The endemic turritellid *Vermicularia fargoi owensi* occurs here on open sand patches between the large foliated clumps of bryozoans.

View of a vermetid worm shell barrier reef off Demijohn Key, Ten Thousand Islands, Collier County, Florida, showing the characteristic reef crest and lagoon geomorphology.

chapter eight

Molluscan faunas of the deep-reef talus macrohabitat (neritic regime)

Introduction

The seaward side of the Florida Keys Reef Tract drops off precipitously, at an acute angle, into the deep waters of the Florida Strait. This plunging, angled slope begins at depths of around 30 m and eventually reaches a terminus in a zone of large stepped ledges and intervening caves at depths of around 100 m. Characteristically, the deep fore-reef area is composed of slabs of coral and masses of coral rubble that have become fused together by sclerosponges and interstitial hexactinellid sponges, producing a stable talus slope. The component coral slabs are the remnants of the reef destruction that takes place during hurricanes and violent storms, when large blocks of coral break off in the heavy surf and tumble down the fore-reef slope. Here, they are cemented together and eventually accumulate enough carbonate rubble to form the substrate for new coral growth. In this way, the reef tract is continuously growing in a seaward direction, with a deterministic growth pattern that eventually will create new reef systems, coral cays, and ultimately, land.

Along the uppermost section of the reef talus slope of the Florida Keys and northward to the Palm Beach coast, the distinctive Pillar Coral, *Dendrogyra cylindrus*, often forms extensive monoculture thickets. Patches of these characteristic deeper-water corals typically interfinger with areas covered by paper-thin, overlapping sheets of Lettuce Corals (*Agaricia* spp.), producing a distinctive deep neritic hermatypic coralline assemblage. Because of the extreme difficulty of collecting on these rough, highly angled slopes that are too deep for scuba diving, little is known about the resident malacofauna. Almost all of the specimens that have been collected from the reef talus slope are taken from deep-water lobster traps that are utilized by commercial lobster fishermen (Figure 8.1). These large wooden traps are baited with rotting meat and then sunk to depths of 100–200 m, where they are left for several months. During their time of submergence, the traps become encrusted with thick growths of fouling organisms such as hydroids, ectoprocts, barnacles, and sessile bivalves such as anomiids, chamids, and pectinids. These artificial miniecosystems create the perfect hiding places for lobsters and support a large array of specialized molluscivorous and carnivorous gastropods. After weeks or months, the traps are pulled to the surface, along with their trapped inhabitants, and scraped clean of all the encrusting organisms. This intensive cleaning process often yields large numbers of rarely seen deep-water mollusks that had taken up residence on the fouling community. Although dependent on commercial fishermen, this collecting technique is still the only efficient and economical way to sample the molluscan faunas of the deep-reef talus zone.

One of the most distinctive gastropods that occurs on the poorly studied deep-reef talus area is the widespread Caribbean-southern Floridian ranellid *Cymatium* (*Ranularia*) *rehderi* (Figure 8.2E, F), which is chosen here to be the namesake for the entire molluscan faunule (the *Cymatium rehderi* Assemblage). This characteristic deep-reef species occurs

Figure 8.1 View of commercial lobster traps stored on Conch Key, Middle Florida Keys. These wooden traps are baited with rotting fish or meat and then submerged in deep-water areas for weeks or months, attracting lobsters and many rare deep-water mollusks. Utilizing lobster traps for collecting shells is the only economical way to sample the molluscan faunas of the Deep-Reef Talus Macrohabitat and the *Cymatium rehderi* Assemblage and *Arca rachelcarsonae* Assemblage. (Photograph by Eddie Matchett.)

Figure 8.2 Mollusks of the *Cymatium rehderi* Assemblage. A = *Marevalvata bairdi* (Dall, 1889), width 5 mm. B = *Laevichlamys multisquamatus* (Dunker, 1864), length 32 mm. C = *Marevalvata tricarinata* (Stearns, 1872), width 4 mm. D = *Spathochlamys benedicti* (Verrill and Bush, 1897), length 12 mm. E, F = *Cymatium (Ranularia) rehderi* Verrill, 1950, length 57 mm. G, H = *Phyllonotus oculatus* (Reeve, 1845), length 83 mm. I = *Chicoreus mergus* E. Vokes, 1974, length 34 mm. J = *Calliostoma fascinans* Schwengel and McGinty, 1942, height 10 mm. K = *Torcula exoleta* (Linnaeus, 1758), length 58 mm. L = *Torculoidella acropora* (Dall, 1889), length 23 mm.

along with a host of interesting and seldom-seen gastropods, such as the calliostomids *Calliostoma fascinans* (Figure 8.2J) and *Calliostoma scalenum* (Figure 8.3D); the beautifully colored turbinid *Turbo (Taenioturbo) cailletii* (Figure 8.3C); the ranellids *Cymatium (Septa) krebsii* (Figure 8.3A) and *Cymatium (Monoplex) mundum*; the harpid *Cancellomorum dennisoni* (Figure 8.3E, F); and the conid *Atlanticonus granulatus* (Figure 8.3J). This last-mentioned species of cone shell (the much-sought-after "Glory of the Atlantic") is a cryptic species, living buried deeply in coral rubble and hexactinellid sponges, and is considered by collectors to be one of the most desirable shells in the Palm Beach and Florida Keys areas. The Glory of the Atlantic lives sympatrically with other cryptic gastropods, such as the liotiids *Marevalvata bairdi* (Figure 8.2A) and *Marevalvata tricarinata* (Figure 8.2C); the muricid *Favartia emipowlusi*; the fasciolariid *Dolicholatirus pauli*; and the drilliid *Fenimorea pagodula*; bivalves such as the pectinids *Laevichlamys multisquamatus* (Figure 8.2B) and *Spathochlamys benedicti* (Figure 8.2D); and the limids *Ctenoides samanensis* and *Ctenoides planulata*. The exposed surface of the coral rubble often forms the substrate for the attachment of exceptionally large and extremely spiny deep-water forms of the spondylid *Spondylus americanus* (Figure 8.3I) and delicately fronded varieties of the deep-water chamid *Chama lactuca*. These sessile bivalves, along with the large anomiid *Pododesmus rudis*, are some of the principal prey items of the large, widespread western Atlantic, deep-water muricid gastropods *Phyllonotus oculatus* (Figure 8.2G, H) and *Chicoreus mergus* (Figure 8.2I).

On more planar areas of the talus slope, large patches of carbonate sand and shell hash often accumulate, forming isolated "islands" of soft substrate on an otherwise-solid carbonate rock substrate. These thick sand patches, which can cover large areas of the angled seafloor, house a number of deep-reef softbottom gastropods, including the suspension-feeding turritellids *Torcula exoleta* (Figure 8.2K) and *Torculoidella acropora* (Figure 8.2L) and the algae-grazing strombid *Aliger gallus* (Figure 8.3G, H). Sand-dwelling zoantherians living in these isolated softbottom areas, along with the adjacent *Dendrogyra* and *Agaricia* coral colonies, serve as the principal prey items for a large number of cnidarian-feeding gastropods such as the epitoniids *Cirsotrema dalli*, *Dentiscala hotessieriana*, and *Epitonium krebsii*, the architectonicids *Heliacus perrieri* and *Philippia krebsii*, and the magilid muricids *Coralliophila aedonis* and *Coralliophila pacei* (Figure 8.3B). The last-mentioned species, *Coralliophila pacei*, with its characteristic two raised bands of scaly flutings, is endemic to the coral coast of southeastern Florida, having only ever been collected on deep reefs off the Palm Beach coast, Pompano Beach (the type locality), and Key Largo.

Biodiversity of the Cymatium rehderi Assemblage

The members of the deep-reef talus slope-associated *Cymatium rehderi* Assemblage are listed here by feeding types and ecological niche preferences. Because this faunule is still poorly known and poorly sampled, the species list that follows can be considered only preliminary and does not accurately reflect the true diversity of the ecosystem. For illustrations of species not illustrated here, see Abbott (1974).

1. HERBIVORES (including algal film grazers)
　Gastropoda
　Fissurellidae
　　Diodora arcuata Sowerby II, 1862
　　Diodora fragilis Farfante and Henriquez, 1947
　　Diodora wetmorei Farfante, 1945

Figure 8.3 Mollusks of the *Cymatium rehderi* Assemblage. A = *Cymatium (Septa) krebsii* Mörch, 1877, length 31 mm. B = *Coralliophila pacei* Petuch, 1987, length 31 mm. This specimen from the deep reefs off Palm Beach is a full adult, showing the characteristic high, turreted spire and two prominent scaly cords. The holotype is a juvenile specimen, only 8 mm in length. C = *Turbo (Taenioturbo) cailletii* Fischer and Bernardi, 1856, width 24 mm. D = *Calliostoma scalenum* Quinn, 1992, height 12 mm. E, F = *Cancellomorum dennisoni* (Reeve, 1842), length 36 mm. G, H = *Aliger gallus* (Linnaeus, 1758), length 123 mm. I = *Spondylus americanus* Hermann, 1781, length 156 mm; deep-water form. J = *Atlanticonus granulatus* (Linnaeus, 1758), length 43 mm. K = *Niveria nix* (Schilder, 1922), length 12 mm.

Lucapina eolis Farfante, 1945
Zeidora bigelowi Perez-Farfante, 1947
Liotiidae
Marevalvata bairdi (Dall, 1889) (Figure 8.2A)
Marevalvata briareus (Dall, 1881)
Marevalvata tricarinata (Stearns, 1872) (Figure 8.2C)
Marevalvata trullata (Dall, 1889)
Turbinidae
Turbo (Taenioturbo) cailletii Fischer and Bernardi, 1856 (Figure 8.3C)
Strombidae
Aliger gallus (Linnaeus, 1758) (Figure 8.3G, H)
2. CARNIVORES
Gastropoda
2a. GENERAL CARNIVORES (including hydroids, sponges, and ectoprocts)
Calliostomatidae
Calliostoma fascinans Schwengel and McGinty, 1942 (Figure 8.2J)
Calliostoma marionae Dall, 1906
Calliostoma scalenum Quinn, 1992 (Figure 8.3D)
Harpidae
Cancellomorum dennisoni (Reeve, 1942) (Figure 8.3E, F)
2b. MOLLUSCIVORES (including drilling predators)
Muricidae (subfamilies Muricinae, Muricopsinae, and Ergalitaxinae; the coralliophagous subfamily Magilinae is listed here separately)
Chicoreus mergus E. Vokes, 1974 (Figure 8.2I)
Favartia emipowlusi (Abbott, 1954)
Phyllonotus oculatus (Reeve, 1845) (Figure 8.2G, H)
Trachypollia sclera Woodring, 1928)
Fasciolariidae
Dolicholatirus pauli (McGinty, 1955)
2c. CNIDARIAN FEEDERS (including corals and zoantherians)
Architectonicidae
Heliacus perrieri (Rochebrune, 1881)
Philippia krebsii (Mörch, 1875)
Epitoniidae
Cirsotrema dalli Rehder, 1945
Cirsotrema pilsbryi (McGinty, 1940)
Dentiscala burryi Clench and Turner, 1950
Dentiscala hotessieriana (d'Orbigny, 1842)
Nodiscala aurifilia (Dall, 1889)
Opaliopsis concava (Dall, 1889) **E**
Muricidae-Magilinae
Coralliophila aedonis (Watson, 1886)
Coralliophila pacei Petuch, 1987 (Figure 8.3B)
2d. ECHINODERM FEEDERS
Ranellidae
Cymatium (Monoplex) mundum (Gould, 1849)
Cymatium (Monoplex) parthenopeum (von Salis, 1793)
Cymatium (Ranularia) rehderi Verrill, 1950 (Figure 8.2E, F)
Cymatium (Septa) krebsii Mörch, 1877 (Figure 8.3A)

2e. VERMIVORES
Conidae-Puncticulinae
Atlanticonus granulatus (Linnaeus, 1758) (Figure 8.3J)
Kellyconus patae (Abbott, 1971)
Conilithidae-Conilithinae
Jaspidiconus mindanus (Hwass, 1792)
Drilliidae
Cerodrillia schroederi Bartsch and Rehder, 1939
Fenimorea pagodula (Dall, 1889)
Raphitomidae
Brachycythara barbarae Lyons, 1972
3. SUSPENSION/FILTER FEEDERS
Gastropoda
Turritellidae
Torcula exoleta (Linnaeus, 1758) (Figure 8.2K)
Torculoidella acropora (Dall, 1889) (Figure 8.2L)
Siliquariidae
Siliquaria squamata Blainville, 1827
Bivalvia
Pectinidae
Laevichlamys multisquamatus (Dunker, 1864) (Figure 8.2B)
Spathochlamys benedicti (Verrill and Bush, 1897) (Figure 8.2D)
Spondylidae
Spondylus americanus Hermann, 1781 (Figure 8.3I)
Limidae
Ctenoides samanensis Stuardo, 1982 (*miumiensis* Mikkelsen and Bieler, 2003, is a synonym)
Ctenoides planulata (Dall, 1886)
Anomiidae
Pododesmus rudis (Broderip, 1834)
Chamidae
Chama lactuca Dall, 1886

A review of the ecological niche preferences of the members of the *Cymatium rehderi* Assemblage shows that specialized carnivorous gastropods dominate this ecosystem.

Molluscan ecology of the deep coral reef ledges

At the base of the fore-reef talus slope, at depths between 100 and 200 m, the accumulations of dead coral slabs create a topography composed of stepped ledges and intervening caves. A large fauna of gorgonian octocorals, composed primarily of antipatharians ("Black Coral"), dominates the surface of these ledges, while the caves shelter a rich fauna of hexactinellid and demospongian poriferans. The shallower upper sections of the ledge and cave zone are often covered with dense growths of deep-water red algae, and these rhodophyte beds, together with the antipatharian thickets and sponge bioherms, house a rich and remarkable molluscan fauna that contains a large number of endemic species. The reef ledge-associated malacofauna extends from the deep reefs off Palm Beach and Broward Counties all the way along the Florida Keys Reef Tract to the area south of the Dry Tortugas. Like the shallower suprajacent reef talus slope, the deep coral ledges are

extremely difficult to sample, and much of the data on the composition of the resident molluscan fauna comes from specimens captured in deep-water lobster traps or dredged by marine research vessels.

The undersides of the ledges provide the substrate for the attachment of an unusual, newly discovered sessile arcid bivalve that appears to be endemic to the deep-water areas off the Florida Keys. This small, stocky, pale-colored ark shell, *Arca rachelcarsonae* (Figure 8.4A–C), was first found attached by byssal threads to a deep-water lobster trap taken off Key Largo and is now known to be a resident of the deep-reef ledges (*Arca rachelcarsonae* n.sp. is described in the Systematic Appendix at the end of this book). This characteristic endemic species is chosen here to represent the entire deep-reef ledge-associated molluscan faunule (the *Arca rachelcarsonae* Assemblage). Several other deep-water bivalves occur with the new arcid and find shelter among the coral ledges, including the pteriid pearl oyster *Pinctada vitrea*; the encrusting oyster *Neopycnodonte cochlear*; the mytilids *Amygdalum politum*, *Amygdalum sagittatum*, and *Crenella decussata*; and the limid *Ctenoides sanctipauli*.

The deep-water red algae beds shelter a large fauna of herbivorous gastropods, including over 19 species in four separate families (the Fissurellidae, Trochidae, Liotiidae, and Modulidae). Of the algivores known from these rhodophyte thickets, 12 belong to the Fissurellidae alone, making this the single largest keyhole limpet fauna known from Florida. Some of the more important and conspicuous fissurellids found in these ledge areas are the large *Diodora (Glyphis) tanneri* and the smaller *Diodora fluviana* and *Nesta atlantica*. These unusual limpets occur together with two species of deep-water modulids: the high-spired, bright pink and purple *Modulus lindae* (Figure 8.4K) and the flattened, star-shaped *Modulus kaicherae* (Figure 8.4M). These two sympatric modulids are the deepest-dwelling species of their family known from anywhere in the western Atlantic. The small trochids *Mirachelus guttarosea*, *Solariella lamellosa*, and *Solariella lacunella* also occur on these deep ledges, along with the liotiids *Sansonia tuberculata* and *Arene variabilis*.

The massively encrusting sponges that line the undersides of the ledges and the interiors of the intervening caves support an interesting and unusual assemblage of spongivorous gastropods. In the deepest areas of the ledges and caves (200-m depths), the pleurotomariid slit shell *Perotrochus amabilis* can be found feeding on encrusting sponges, along with the deep-water abalone *Haliotis pourtalesi*. Several obligate spongivores also occur with the slit shell and abalone, including the rare cowrie shell *Propustularia surinamensis* (Figure 8.4H, I) and the large calliostomatids *Calliostoma (Kombologion) benedicti* (Figure 8.4D), *Calliostoma*

Figure 8.4 Mollusks of the *Arca rachelcarsonae* Assemblage. A = *Arca rachelcarsonae* Petuch and Myers, new species; holotype, length 16 mm, lateral view. B = *Arca rachelcarsonae* Petuch and Myers, new species; holotype, length 16 mm, view of the ligamental shelf (see Systematic Appendix). C = *Arca rachelcarsonae* Petuch and Myers, new species. Paratype, length 21 mm, lateral view. D = *Calliostoma benedicti* Dall, 1889, width 28 mm. E = *Bursa (Colubrellina) ranelloides tenuisculpta* (Dautzenberg and Fischer, 1906), length 42 mm (*B. finlayi* is a synonym). F, G = *Pusula juyingae* Petuch and Sargent, 2011, length 14 mm. This large endemic triviid has been incorrectly referred to as *Pusula costispunctata* by several Florida workers. That taxon is actually an Eastern Pacific Panamic Molluscan Province species. H, I = *Propustularia surinamensis* (Perry, 1811), length 26.8 mm. Four specimens of this rare cowrie were collected alive on the deep reefs off Palm Beach by Thomas and Paula Honker in the 1980s. Another specimen is also known to have been collected from lobster traps placed on deep-water reefs off northern Key Largo. These southeastern Floridian specimens mark the northernmost end of the range of this rare and desirable western Atlantic cowrie. J = *Pseudocyphoma gibbulum* (Cate, 1978), length 13.5 mm. K = *Modulus lindae* Petuch, 1987, length 11 mm. L = *Attiliosa philippiana* (Dall, 1889), length 16 mm. M = *Modulus kaicherae* Petuch, 1987, view of spire, width 12 mm.

(*Kombologion*) *hendersoni,* and *Calliostoma* (*Kombologion*) *sayanum.* Several types of encrusting botrylid colonial tunicates (Urochordata) are interspersed between the sponge patches, and these provide both the food source and the substrate for the rare triviid cowries *Niveria nix* (Figure 8.3K) and *Pusula juyingae* (Figure 8.4F, G). To date, this last-mentioned triviid has only been collected along the deep reefs off Palm Beach and Broward Counties, but it undoubtedly ranges southward to at least the northern Keys.

Of all the resident gastropod groups in the *Arca rachelcarsonae* Assemblage, the Ovulidae exhibits the highest level of endemism, with virtually every species restricted to the area extending from Palm Beach to the Dry Tortugas. Here, at least four species of endemic egg shells live on the antipatharian octocorals, with the most frequently encountered being *Pseudocyphoma gibbulum* (Figure 8.4J), *Pseudosimnia vanhyningi*, *Pseudosimnia sphoni*, and *Adamantia solemi*. These octocorals also provide the substrate for the cnidarian-feeding pediculariid *Pedicularia decussata*, where the variably shaped shell adheres to the stem or holdfast of the antipatharian colony. The pediculariid conforms to the shape of the irregular substrate, so no two individuals will have the same exact shell form. Besides the antipatharians, other cnidarians, such as large zoantherian colonies, also encrust the deep-reef ledges, and these are the principal prey items of the cnidarian-feeding magiline muricids *Coralliophila richardi* and *Babelomurex dalli*.

The encrusting bivalve fauna of the deep ledges, including species such as *Arca rachelcarsonae* and *Neopycnodonte cochlear*, provide the main food resources for the drilling molluscivores of the family Muricidae. A particularly rich assemblage of muricids is associated with the deep coral ledges, including rarely seen species such as *Pterochelus ariomus*, *Trossulasalpinx macra*, *Pteropurpura bequaerti*, *Pterynotus guesti*, *Pterynotus phaneus*, *Attiliosa philippiana* (Figure 8.4L), *Vokesimurex morrisoni* (Figure 8.5K), *Favartia goldbergi* (Figure 8.5A), *Dermomurex* (*Trialatella*) *glicksteini* (Figure 8.5E), and *Pygmaepterys richardbinghami* (Figure 8.5D). The last three of these muricid species were originally collected on the deep reefs off Palm Beach, but future research will undoubtedly show that they occur throughout the Florida Keys in the right habitat type and depth range. These deep-water murex shells also live along with the small echinoderm-feeding ranellid *Cymatium* (*Septa*) *pharcidum* and the sipunculid and nemertean-feeding bursids *Bursa* (*Colubrellina*) *ranelloides tenuisculpta* (Figure 8.4E) and *Bufonaria bufo*. The rare plesiotritonine cancellariid *Tritonoharpa janowskyi* (Figure 8.5C) also occurs in this habitat and is probably a suctorial feeder on polychaete and sipunculid worms. Although originally described from the deep reefs off Palm Beach, this stocky, truncated *Tritonoharpa* species has also been collected from deep lobster traps brought up from off northern Key Largo.

Figure 8.5 Gastropods of the *Arca rachelcarsonae* Assemblage. A = *Favartia goldbergi* Petuch and Sargent, 2011; holotype, length 10.5 mm. B = *Gemophos filistriatus* Vermeij, 2006, length 20.4 mm. C = *Tritonoharpa janowskyi* Petuch and Sargent, 2011; holotype, length 15.8 mm. D = *Pygmaepterys richardbinghami* (Petuch, 1987); holotype, length 16 mm. E = *Dermomurex* (*Trialatella*) *glicksteini* Petuch, 1987; holotype, length 16 mm. F, G = *Dauciconus amphiurgus* (Dall, 1889), length 26.5 mm. H = *Kellyconus binghamae* (Petuch, 1987); holotype, length 17.4 mm. I = *Tuckericonus flamingo* (Petuch, 1980); holotype, length 19.3 mm. J = *Dauciconus glicksteini* (Petuch, 1987); holotype, length 20.7 mm. K = *Vokesimurex morrisoni* Petuch and Sargent, 2011, length 41 mm. This deep-water species has been incorrectly referred to as *Vokesimurex bellegladeensis* (E. Vokes, 1963) by many workers in Florida. The true *bellegladeensis* is actually an early and middle Pleistocene fossil from the Bermont Formation of southern Florida (see Figure 1.5B, C in Chapter 1). The 1.5-million-year-old fossil species differs from the living *morrisoni* in having a much thicker and heavier shell, in having proportionally thicker and wider varices, and in having coarser shell sculpture.

An unusually large biomass of interstitial polychaete worms must be present on the sponge-covered ledges and caves of the deep fore reef, as attested by the exceptionally large and diverse fauna of vermivorous gastropods in the toxoglossate families Conidae, Drilliidae, and Raphitomidae. The brightly colored cone shells *Dauciconus amphiurgus* (Figure 8.5F, G), *Dauciconus glicksteini* (Figure 8.5J), *Gradiconus patglicksteinae*, *Kellyconus binghamae* (Figure 8.5H), and *Tuckericonus flamingo* (Figure 8.5I) occur in sand patches scattered between the coral ledges and in areas of mixed sand and coral rubble. With the exception of the wide-ranging Carolinian Province indicator species *Dauciconus amphiurgus*, the other three cones are restricted to the Palm Beach and Broward coasts and to the deep-reef areas off the Florida Keys and Dry Tortugas (see Tucker, 2013, for an overview of these poorly known deep-water cones).

With 20 resident species in seven genera, the family Raphitomidae is the most diverse group of vermivores found in the deep terrace area. Some of the most frequently encountered species include *Mangelia rugirima*, *Kurtziella limonitella*, *Brachycythara biconica*, *Cryoturris cerinella*, *Cryoturris quadrilineata*, *Daphnella lymnaeiformis*, *Daphnella stegeri*, and *Eubela macgintyi* (all 20 species are listed in the next section). The family Drilliidae is also an important group of vermivores in the *Arca rachelcarsonae* Assemblage, containing such well-known deep-water species as *Cerodrillia hendersoni*, *Leptadrillia splendida*, and *Bellaspira pentagonalis*. The turroidean toxoglossate fauna of the deep ledges environment also contains at least nine species of the horaiclavid genus *Inodrillia* (listed in the next section). This endemic radiation, described by Paul Bartsch of the Smithsonian Institution in 1943, is one of the largest turroidean species swarms of a single genus ever found anywhere in Florida.

Biodiversity of the Arca rachelcarsonae Assemblage

The members of the deep coral reef ledge-associated *Arca rachelcarsonae* Assemblage are listed here by feeding type and ecological niche preferences. Because of the difficulty in conducting biological sampling in this rough, rocky environment, the resident malacofauna is still poorly known. The following list must be considered only preliminary, and future research will doubtlessly bring to light many more new and undescribed species. For illustrations of many of the species not shown here, see the iconography of Abbott (1974).

1. **HERBIVORES (including algal scrapers and red algae feeders)**
   Gastropoda
   Fissurellidae (some of these taxa may also feed on sponges)
       Diodora fluviana Dall, 1889
       Diodora (Glyphis) tanneri Verrill, 1883
       Nesta atlantica Farfante, 1947
       Puncturella acuminata Watson, 1883
       Puncturella agger Watson, 1883
       Puncturella billsae Farfante, 1947
       Puncturella erecta Dall, 1889
       Puncturella granulata Seguenza, 1863
       Rimula aequisculpta Dall, 1927
       Rimula dorriae Farfante, 1947
       Rimula frenulata Dall, 1889
       Rimula pycnonema Pilsbry, 1943

Trochidae
> *Mirachelus guttarosea* Dall, 1889
> *Solariella lacunella* (Dall, 1881)
> *Solariella lamellosa* Verrill and Smith, 1880

Liotiidae
> *Marevalvata variabilis* (Dall, 1886)
> *Sansonia tuberculata* (Watson, 1886)

Modulidae
> *Modulus kaicherae* Petuch, 1987 (Figure 8.4M)
> *Modulus lindae* Petuch, 1987 (Figure 8.4K)

2. CARNIVORES
Gastropoda

2a. SPONGIVORES
Pleurotomariidae
> *Perotrochus amabilis* (F. Bayer, 1963)

Haliotidae
> *Haliotis pourtalesi* Dall, 1881

2b. GENERAL CARNIVORES/SCAVENGERS (including sponges and hydroids)
Calliostomatidae
> *Calliostoma (Kombologion) benedicti* Dall, 1889 (Figure 8.4D)
> *Calliostoma (Kombologion) hendersoni* Dall, 1927
> *Calliostoma (Kombologion) psyche* Dall, 1889
> *Calliostoma (Kombologion) sayanum* Dall, 1889

Mathildidae (presumed spongivore/grazing carnivore; ecology still problematical)
> *Mathilda barbadensis* Dall, 1889
> *Mathilda hendersoni* Dall, 1927
> *Mathilda yucatecana* Dall, 1927

Cypraeidae
> *Propustularia surinamensis* (Perry, 1811) (Figure 8.4H, I)

Buccinidae
> *Antillophos candeanus* (d'Orbigny, 1842) (in sand pockets on the deep reef)
> *Parviphos adelus* (Schwengel, 1942)

2c. PROBLEMATICAL (suctorial feeders?)
Cancellariidae
> *Tritonoharpa janowskyi* Petuch and Sargent, 2011 (Figure 8.5C)
> *Ventrilia tenerum* (Philippi, 1848)

2d. UROCHORDATE FEEDERS (colonial tunicates)
Triviidae
> *Niveria nix* (Schilder, 1922) (Figure 8.3K)
> *Pusula juyingae* Petuch and Sargent, 2011 (Figure 8.4F, G) [This endemic southeastern Florida triviid has been incorrectly referred to as *Pusula costispunctata* (Sowerby II, 1870) by several recent workers; that taxon is actually a Panamic Molluscan Province species from the eastern Pacific.]

2e. CNIDARIAN FEEDERS (antipatharians, zoantherians, scleractinians)
Ovulidae
> *Adamantia solemi* Cate, 1973
> *Pseudocyphoma aureocincta* (Dall, 1889)
> *Pseudocyphoma gibbulum* (Cate, 1978) (Figure 8.4J)

Pseudosimnia sphoni Cate, 1973
Pseudosimnia vanhyningi (M. Smith, 1940)
Pediculariidae
Pedicularia decussata (Gould, 1855)
Muricidae-Magilidae (scleractinians and zoantherians)
Babelomurex dalli (Emerson and Puffer, 1963)
Coralliophila richardi (Fischer, 1882) (= *C. lactuca*)
2f. MOLLUSCIVORES (including drilling predators)
Muricidae-Muricinae, Muricopsinae, Ocenebrinae, Typhinae
Attiliosa philippiana (Dall, 1889) (Figure 8.4L)
Dermomurex (Trialatella) glicksteini Petuch, 1987 (Figure 8.5E)
Favartia goldbergi Petuch and Sargent, 2011 (Figure 8.5A)
Pterochelus ariomus (Clench and Farfante, 1945)
Pteropurpura bequaerti (Clench and Farfante, 1945)
Pterynotus guesti Harasewych and Jensen, 1979.
Pterynotus phaneus (Dall, 1889)
Pygmaepterys richardbinghami (Petuch, 1987) (Figure 8.5D)
Siratus consuela (Verrill, 1950) (off the Dry Tortugas)
Trossulasalpinx macra (Verrill, 1887)
Typhinellus sowerbii (Broderip, 1833)
Vokesimurex morrisoni Petuch and Sargent, 2011 (Figure 8.5K)
Fasciolariidae
Dolicholatirus pauli (McGinty, 1955)
Buccinidae
Antillophos beaui (Fischer and Bernardi, 1857)
Bailya weberi (Watters, 1983)
Gemophos filistriatus Vermeij, 2006 (Figure 8.5B)
Hesperisternia karinae (Usticke, 1959)
2g. ECHINODERM FEEDERS (asteroids, ophiuroids, holothurians)
Ranellidae
Cymatium (Septa) pharcidum Dall, 1889
2h. VERMIVORES (including polychaetes, sipunculids, and nemerteans)
Bursidae
Bufonaria bufo (Bruguiere, 1792)
Bursa (Colubrellina) ranelloides tenuisculpta (Dautzenberg and Fischer, 1906)
(Figure 8.4E)
Mitridae
Subcancilla straminea (A. Adams, 1853)
Conidae-Puncticulinae
Dauciconus amphiurgus (Dall, 1889) (Figure 8.5F, G)
Dauciconus glicksteini (Petuch, 1987) (Figure 8.5J)
Gradiconus patglicksteinae (Petuch, 1987)
Kellyconus binghamae (Petuch, 1987) (Figure 8.5H)
Tuckericonus flamingo (Petuch, 1980) (Figure 8.5I)
Drilliidae
Bellaspira pentagonalis (Dall, 1889)
Cerodrillia hendersoni Bartsch, 1943

    Cerodrillia williami Bartsch, 1943
    Drillia acrybia (Dall, 1889)
    Drillia pharcida (Dall, 1889)
    Leptodrillia splendida (Bartsch, 1934)

Horaiclavidae
    Inodrillia acova Bartsch, 1943
    Inodrillia avira Bartsch, 1943
    Inodrillia dido Bartsch, 1943
    Inodrillia gibba Bartsch, 1943
    Inodrillia hesperia Bartsch, 1943
    Inodrillia hilda Bartsch, 1943
    Inodrillia ino Bartsch, 1943
    Inodrillia miamia Bartsch, 1943
    Inodrillia vetula Bartsch, 1943

Zonulispiridae
    Compsodrillia disticha Bartsch, 1934
    Compsodrillia eucosmia (Dall, 1889)
    Compsodrillia haliostrephis (Dall, 1889)

Clathurellidae
    Lioglyphostoma adematum Woodring, 1928
    Nannodiella vespuciana (d'Orbigny, 1847)

Raphitomidae
    Brachycythara biconica (C.B. Adams, 1850)
    Cryoturris cerinella (Dall, 1889)
    Cryoturris filifera (Dall, 1881)
    Cryoturris quadrilineata (C.B. Adams, 1850)
    Daphnella corbicula Dall, 1889
    Daphnella elata Dall, 1889
    Daphnella lymnaeiformis (Kiener, 1840)
    Daphnella retifera Dall, 1889
    Daphnella stegeri McGinty, 1955
    Eubela mcgintyi Schwengel, 1943
    Kurtziella accincta (Montagu, 1808)
    Kurtziella atrostyla (Tryon, 1884)
    Kurtziella limonitella (Dall, 1883)
    Kurtziella perryae Bartsch and Rehder, 1939
    Kurtziella serga (Dall, 1881)
    Pyrgocythara balteata (Reeve, 1846)
    Pyrgocythara densestriata (C.B. Adams, 1850)
    Rimosodaphnella morra (Dall, 1881)
    Rubellatoma diomedea Bartsch and Rehder, 1939
    Rubellatoma rubella (Kurtz and Stimpson, 1851)

3. SUSPENSION/FILTER FEEDERS
Bivalvia
Arcidae
    Arca rachelcarsonae Petuch and Myers, new species (Figure 8.4A–C) (see
      Systematic Appendix)

Mytilidae
>   *Amygdalum politum* (Verrill and Smith, 1880)
>   *Amygdalum sagittatum* (Rehder, 1935)
>   *Crenella decussata* (Montagu, 1808)

Pteriidae
>   *Pteria vitrea* (Reeve, 1857)

Gryphaeidae
>   *Neopycnodonte cochlear* (Poli, 1795)

Limidae
>   *Ctenoides sanctipauli* Stuardo, 1982

A review of the ecological niche preferences of the members of the *Arca rachelcarsonae* Assemblage shows that the specialized carnivorous gastropods dominate the ecosystem by more than a three-to-one ratio. The biomass of the cnidarian, polychaete, sipunculid, urochordate, and poriferan prey items appear to surpass that of any other known marine ecosystem in the Florida Keys.

chapter nine

Molluscan faunas of deep softbottom and pelagic macrohabitats (oceanic regime)

Introduction

The Oceanic Regime of the Florida Keys, Dry Tortugas, and coasts of Broward and Palm Beach Counties encompasses all the peripheral areas that are under more than 200 m of seawater. Some of these adjacent areas, which extend seaward from the edge of the shallow continental shelf, plunge to depths of over 400 m, making them the deepest ecosystems classified within the Coastal Marine Ecological Classification Standard (CMECS) system. The Florida Keys Oceanic Regime actually takes into account three separate macrohabitats that reside within different bathymetric ranges. The first of these, the Deep Coralline Algal Beds, occurs in depths of over 200 m along southwestern Florida and the areas north of the Dry Tortugas, at the mouth of the South Florida Bight. In contrast, the second macrohabitat, the Deep Softbottom Terraces, extends from the Palm Beaches all the way to the Marquesas Keys and occurs in much deeper areas, with depths ranging from 200 to 480 m. The third macrohabitat, the Open Oceanic Pleuston, represents all the molluscan communities that float on the offshore sea surface, well above the abyssal seafloor of the deep Florida Straits. The molluscan faunas of each of these disparate deep-water environments are still poorly studied and may contain many new, undescribed species.

Molluscan ecology of the deep coralline algal beds

The coralline algal beds of the Florida Keys and southwestern Florida form a narrow band that extends from the Pulley Ridge, 160 km west of the Dry Tortugas, northward along the edge of the Florida continental shelf to at least the same latitude as Sarasota. The north-south-oriented Pulley Ridge itself is over 100 km in length and ranges, bathymetrically, from 60 to over 200 m deep. The lower sections of the ridge, from around 80 to 100 m deep, house the deepest hermatypic coral reefs in the continental United States, composed almost entirely of blue and purple corals of the genera *Agaricia* and *Leptoceris*. These deep reefs are nourished by the extremely warm and nutrient-rich waters of the Tortugas Gyre, a large eddy current that is permanently established over the western edge of the South Florida Bight. The crystal clear waters of the Tortugas Gyre allow light to penetrate to its maximum depth before becoming attenuated, creating the perfect conditions for deep-water red algal growth.

Below 100 m, the *Agaricia* and *Leptoceris* purple coral reefs of the Pulley Ridge grade directly into extensive beds of large, rounded rhodoliths (red algal nodules), produced by the deep-water coralline rhodophyte alga *Porolithon*. At many localities, these rounded growth forms interfinger with dense aggregations of small branching growth forms, accreting together to create a dense porous mass that is often over 10 m thick (Petuch and Sargent, 2011c: 90–93). North of the Pulley Ridge and along the edge of the western Florida continental shelf, in particular, these intertwined rhodolith beds are especially well developed, forming a seafloor substrate that is unique in the southeastern United States. Because of their remoteness from the Floridian Peninsula and because of the excessive depths, the ecosystems of the rhodolith beds near the Dry Tortugas are rarely sampled and are essentially unexplored. As in the case of the deep reefs off the Florida Keys, lobster traps are regularly employed within this area by commercial fishermen, and these are often the only way to easily sample the resident molluscan fauna.

The small amount of data that has been gleaned from lobster trap collections and dredging expeditions on research vessels has shown that the deep coralline algae beds shelter a highly unusual molluscan fauna containing many still-undescribed species. One of the most distinctive of the rhodolith-associated gastropods, the muricid *Chicoreus rachelcarsonae* (Figure 9.1A, B), is endemic to this unusual macrohabitat and has been selected here to represent the entire molluscan faunule (the *Chicoreus rachelcarsonae* Assemblage). This delicate and beautiful muricid, which differs from the closely related shallow-water *Chicoreus dilectus* by having only five spines on its body whorl instead of six and by having a pink or salmon-colored shell, occurs together with another endemic rhodolith muricid, *Phyllonotus whymani* (Figure 9.1D, E). Both of these iconic carnivorous gastropods feed by drilling the shells of several larger bivalves, including the sessile-attached chamids *Chama lactuca* Dall, 1886 and *Chama sinuosa* Broderip, 1835; the brightly colored spiny scallop *Lindapecten lindae* (Figure 9.2G, H); the cardiid *Americardia lightbourni* (Figure 9.3B); and the venerids *Lirophora paphia* (Linnaeus, 1758) and *Lirophora varicosa* (Sowerby II, 1853) (Figure 9.2K; often incorrectly referred to as *Lirophora latilirata*, which is actually a Pliocene fossil species). A host of smaller and more delicate deep-water muricids also live among the finely branched rhodolith matrix, and these include *Favartia lindae* (Figure 9.1F), *Murexiella kalafuti* (Figure 9.1G), *Murexiella levicula* (Figure 9.1I), *Vokesimurex lindajoyceae* (Figure 9.1J), and *Murexiella taylorae* (Figure 9.2J). Smaller, shallowly infaunal bivalves such as the tellinids *Strigilla pisiformis* and *Cymatoica hendersoni* and the cardiid *Microcardium peramabile* (Dall, 1881) serve as the principal prey items of these drilling predators.

The dense rhodolith aggregations also serve as the attachment areas for large antipatharian gorgonian colonies and sponges, increasing the number of demes and microhabitats available to the resident malacofauna. These gorgonians serve as both the food resource and the substrate of the rarely seen ovulid gastropods *Cyphoma lindae* (Figure 9.1C) and *Cyphoma robustior*, both deep-water species found from the Dry Tortugas northward to off Apalachicola. Several deep-water spongivorous gastropods also live on the rhodolith demosponges, including the calliostomatid *Calliostoma sapidum* and the sponge-dwelling abalone *Haliotis pourtalesi* (Figure 9.1M). Several rarely seen and poorly known herbivorous gastropods also live on the coralline algal seafloors, feeding directly on the red algal nodules. Some of the more prominent of these include the liotiids *Cyclostrema huesonicum* and *Cyclostrema tortuganum* and the large cerithiids *Cerithium chara* and *Cerithium lymani*. These unusual deep-water herbivores occur along with several prominent suspension-feeding gastropods, including the heavily sculptured turritellid *Torculoidella lindae* (Figure 9.1H) and the siliquariids *Siliquaria modesta* and *Siliquaria squamata*.

A large and highly endemic fauna of mollusk-feeding fasciolariid gastropods also occurs along with the muricids on the rhodolith seafloor, including the endemic Tortugas *Leucozonia jacarusoi* (Figure 9.1K) and *Cinctura tortuganum* (Figure 9.1L) and more widespread species such as *Fasciolaria bullisi* (Figure 9.3A), *Heilprinia lindae* (Figure 9.2E; restricted to the Dry Tortugas and western Florida), and *Fusinus stegeri*. Of the endemic

Tortugas species, *Leucozonia jacarusoi* is one of the most interesting, as it has only ever been collected in deep-water lobster traps, where it was found feeding on encrusting barnacles and small sessile bivalves. Although often incorrectly synonymized with the common shallow-water *Leucozonia nassa*, this deep-water Tortugas endemic differs in that it has a smaller, more elongated shell with a much longer siphonal canal, in having a stronger corded shell sculpture, and in always being either pure white or pale yellow in color. The large carnivorous fasciolariid gastropods occur along with several potential competitors, such as the thin-shelled deep-water busyconid *Lindafulgur lyonsi* (Figure 9.3G); the small buccinids *Hesperisternia harasewychi* (Figure 9.3C, D) and *Hesperisternia sulzyckii* (Figure 9.3E, F; see Systematic Appendix); the small scaphelline volute *Caricellopsis matchetti* (Figure 9.3H, I); and the olivid *Americoliva sunderlandi* (Figure 9.3J, K). One of the more ecologically problematic species found in the *Chicoreus rachelcarsonae* Assemblage is the large, high-spired cancellariid *Cancellaria richardpetiti* (Figure 9.3L). This rare species may feed suctorially on sleeping elasmobranchs such as stingrays, piercing the fish's gills with a needle-like tooth and then pumping the blood out of its prey through a long, extendible proboscis. Closely related cancellariid species from California have been observed to feed in this manner, and the deep-water southwestern Florida species may also be a similar suctorial feeder.

The *Chicoreus rachelcarsonae* Assemblage also houses a large and complex fauna of vermivorous gastropods, comprising six families and nine genera. The most prominent of these worm-eating species are the conids *Dauciconus amphiurgus*, *Dauciconus aureonimbosus* (Figure 9.2A; endemic to the Dry Tortugas and the Pulley Ridge area), *Gradiconus philippii* (Figure 9.2B; see Tucker, 2013: 117–121), and *Lindaconus aureofasciatus* (Figure 9.2D), and the conilithid *Kohniconus delessertii* (Figure 9.2F), which often occur in large numbers on and within the rhodolith substrate. The cone shells live together with other large and conspicuous vermivores, such as the terebrids *Myurellina lindae* (Figure 9.2C; endemic to the Dry Tortugas and southwestern Florida) and *Myurellina floridana*; the drilliids *Fenimorea sunderlandi* (Figure 9.2I), *Fenimorea fucata* (Figure 9.3M), *Neodrillia brunnescens*, and *Neodrillia woodringi*; and the turrid *Polystira sunderlandi* (Figure 9.2L). The species richness of vermivores attests to the presumed large biomass and diversity of the resident polychaete, sipunculid, and nemertean worm fauna, most of which live interstitially between the algal rhodoliths.

Biodiversity of the Chicoreus rachelcarsonae *Assemblage*

The members of the deep-water red coralline algae-associated *Chicoreus rachelcarsonae* Assemblage are listed here by feeding type and ecological niche preferences. This list must be considered only preliminary, as future research and collecting will undoubtedly bring to light many more species that are new and undescribed. For illustrations of taxa not shown here, see the iconography of Abbott (1974; for gastropods) and Mikkelsen and Bieler (2008; for bivalves). Several of these gastropods and bivalves also occur in the deeper softbottom habitats of the Hawk Channel (the *Polinices lacteus* Assemblage) and in sand patches interspersed between deep talus slope reef corals (the *Cymatium rehderi* Assemblage).

1. HERBIVORES (including red algal grazers)
Gastropoda
Liotiidae
    Cyclostrema huesonicum Dall, 1927
    Cyclostrema tortuganum (Dall, 1927)

Cerithiidae
> *Cerithium chara* Pilsbry, 1949
> *Cerithium lymani* Pilsbry, 1949

Strombidae
> *Strombus alatus* Gmelin, 1791 (spineless, heavily grooved, deep-water dwarf form)

Xenophoridae
> *Xenophora microdiscus* Petuch, 1994 (Figure 9.2M) (This small, very flattened species was described originally as a late Pleistocene fossil from southern Florida but is now known to be living on coralline algae beds off western Florida and the Dry Tortugas.)

2. CARNIVORES
Gastropoda

2a. GENERAL CARNIVORES/SCAVENGERS
Cypraeidae
> *Erosaria acicularis* (Gmelin, 1791) (heavily enameled deep-water dwarf form)
> *Macrocypraea (Lorenzicypraea) cervus* (Linnaeus, 1771) (giant deep-water form; often over 140 mm in length)

Personidae (feeding preferences problematical; possibly polychaete worms)
> *Distorsio clathrata* (Lamarck, 1816)
> *Distorsio mcgintyi* Emerson and Puffer, 1953

Buccinidae
> *Hesperisternia harasewychi* (Petuch, 1987) (Figure 9.3C, D)
> *Hesperisternia sulzyckii* Petuch and Myers, new species (Figure 9.4D, E; see Systematic Appendix)

Volutidae
> *Caricellopsis matchetti* Petuch and Sargent, 2011 (Figure 9.3H, I)
> *Scaphella junonia elizabethae* Petuch and Sargent, 2011 (large Tortugas form resembling the subspecies *butleri* from the Campeche Banks of Mexico)

Olividae
> *Americoliva sunderlandi* (Petuch, 1987) (Figure 9.3J, K)

Figure 9.1 Gastropods of the *Chicoreus rachelcarsonae* Assemblage. A, B = *Chicoreus rachelcarsonae* Petuch, 1987, length 40 mm. This classic deep-water coralline algal bed species has been incorrectly synonymized with the common shallow-water *Chicoreus dilectus*, but differs in having five varical spines instead of six, in having a proportionally broader and wider body whorl and varices, in having a proportionally larger protoconch, and in having a pink, orange, or salmon-pink shell color. C = *Cyphoma lindae* Petuch, 1987; holotype, length 17 mm. Shell is always pale tan or brown; mantle of living animal colored with very numerous tiny black spots. D, E = *Phyllonotus whymani* Petuch and Sargent, 2011; holotype, length 43.4 mm. This deep-water species differs from the closely related shallow-water *Phyllonotus pomum* in that it has a smaller, stockier shell with a proportionally lower spire; in having thinner, slightly recurved varices; in having stronger and more prominent spiral cords; in having a smaller and less-developed parietal shield; in lacking any scales, spines, or flutings on the varices and siphonal canal; and in having shell colors of only pink, salmon, or orange. F = *Favartia lindae* Petuch, 1987; holotype, length 18 mm. G = *Murexiella kalafuti* Petuch, 1987; holotype, length 15.4 mm. H = *Torculoidella lindae* (Petuch, 1987); holotype, length 22.5 mm. I = *Murexiella levicula* (Dall, 1889), length 18 mm. J = *Vokesimurex lindajoyceae* (Petuch, 1987), length 32.4 mm. K = *Leucozonia jacarusoi* Petuch, 1987; holotype, length 24 mm. Shell color varies from pure white to pale yellow. L = *Cinctura tortugana* (Hollister, 1957), length 70.8 mm. M = *Haliotis pourtalesi* Dall, 1881, length 17 mm.

Marginellidae
> *Eratoidea hematita* (Kiener, 1843)
> *Prunum hartleyana* (Schwengel, 1941)

2b. PROBLEMATICAL (suctorial feeder on stingrays and elasmobranchs?)

Cancellariidae
> *Cancellaria richardpetiti* Petuch, 1987 (Figure 9.3L)

2c. SPONGIVORES

Haliotidae
> *Haliotis pourtalesi* Dall, 1881 (Figure 9.1M)

Calliostomatidae
> *Calliostoma sapidum* Dall, 1881

2d. CNIDARIAN FEEDERS (including gorgonians and antipatharians)

Ovulidae
> *Cyphoma lindae* Petuch, 1987 (Figure 9.1C)
> *Cyphoma robustior* Bayer, 1941

2e. MOLLUSCIVORES (including drilling predators)

Muricidae (Muricinae, Muricopsinae)
> *Chicoreus rachelcarsonae* Petuch, 1987 (Figure 9.1A, B)
> *Favartia lindae* Petuch, 1987 (Figure 9.1F)
> *Hexaplex fulvescens* (Sowerby I, 1834)
> *Murexiella kalafuti* Petuch, 1987 (Figure 9.1G)
> *Murexiella levicula* (Dall, 1889) (Figure 9.1I)
> *Murexiella taylorae* Petuch, 1987 (Figure 9.2J)
> *Phyllonotus whymani* Petuch and Sargent, 2011 (Figure 9.1D, E)
> *Siratus consuela* (Verrill, 1950)
> *Vokesimurex lindajoyceae* (Petuch, 1987) (Figure 9.1J)
> *Vokesimurex morrisoni* Petuch and Sargent, 2011

Fasciolariidae
> *Cinctura tortugana* (Hollister, 1957) (Figure 9.1L)

Figure 9.2 Mollusks of the *Chicoreus rachelcarsonae* Assemblage. A = *Dauciconus aureonimbosus* (Petuch, 1987); holotype, length 25.5 mm. B = *Gradiconus philippii* (Kiener, 1847), length 25.4 mm. Note pink interior. C = *Myurellina lindae* (Petuch, 1987); holotype, length 63 mm. D = *Lindaconus aureofasciatus* (Rehder and Abbott, 1951); holotype, length 66.6 mm. Although generally considered to be a deep-water variant of the common shallow-water *Lindaconus atlanticus*, this deep-water shell appears to be a distinct and separate species. E = *Heilprinia lindae* Petuch, 1987; holotype, length 141 mm. Although sometimes synonymized with the common shallower-water *Heilprinia timessus*, this deep-water western Florida shell differs in that it is much more elongated, with a far more protracted spire; in having a thin, distinct, deeply incised channel along the sutures of the spire whorls; and in having a pure white or pinkish-white shell color. F = *Kohniconus delessertii* (Recluz, 1843), length 52 mm. G, H = *Lindapecten lindae* Petuch, 1995; holotype, length 16 mm. I = *Fenimorea sunderlandi* (Petuch, 1987); holotype, length 39.7 mm. J = *Murexiella taylorae* Petuch, 1987; holotype, length 15.6 mm. K = *Lirophora varicosa* (Sowerby II, 1853), width 24 mm. This heavily ribbed offshore species is often incorrectly referred to as *Chione latilirata* Conrad, which is actually an early Pliocene fossil species from Virginia and North Carolina. L = *Polystira sunderlandi* Petuch, 1987; holotype, length 27.7 mm. M = *Xenophora microdiscus* Petuch, 1994, width 15 mm. Originally named as a Pleistocene fossil, this small, flattened species is still living in deep water off western Florida and the Dry Tortugas. Although similar to the much larger, shallow-water *Xenophora conchyliophora*, this relictual species has a smooth shell base and lacks the pebbly surface seen on the shell base of its better-known congener.

Figure 9.3 Mollusks of the *Chicoreus rachelcarsonae* Assemblage. A = *Fasciolaria bullisi* Lyons, 1972; holotype, length 134.2 mm. B = *Americardia lightbourni* Lee and Huber, 2012, length 11 mm. C, D = *Hesperisternia harasewychi* (Petuch, 1987); holotype, length 24.8 mm. E, F = *Hesperisternia sulzyckii* Petuch and Myers, new species; holotype, length 26.3 mm (see Systematic Appendix for the description). G = *Lindafulgur lyonsi* (Petuch, 1987); holotype, length 128 mm. H, I = *Caricellopsis matchetti* (Petuch and Sargent, 2011); holotype, length 35.9 mm. Note the strong, coarse reticulate shell sculpture and rounded, domelike calcarella. J, K = *Americoliva sunderlandi* (Petuch, 1987); holotype, length 22 mm. L = *Cancellaria richardpetiti* Petuch, 1987, length 47 mm. M = *Fenimorea fucata* (Reeve, 1845), length 22 mm.

      Fasciolaria bullisi Lyons, 1972 (Figure 9.3A)
      Fusinus stegeri Lyons, 1978
      Heilprinia lindae Petuch, 1987 (Figure 9.2E)
      Heilprinia timessus (Dall, 1889)
      Leucozonia jacarusoi Petuch, 1987 (Figure 9.1K)

Busyconidae
      Fulguropsis plagosum (Conrad, 1863)
      Lindafulgur lyonsi (Petuch, 1987) (Figure 9.3G)

2f. VERMIVORES (including polychaetes, sipunculids, and nemerteans)

Conidae-Puncticulinae
      Dauciconus amphiurgus (Dall, 1889)
      Dauciconus aureonimbosus (Petuch, 1987) (Figure 9.2A)
      Gradiconus philippii (Kiener, 1845) (Figure 9.2B)
      Lindaconus aureofasciatus (Rehder and Abbott, 1951) (Figure 9.2D)

Conilithidae-Conilithinae
      Kohniconus delessertii (Recluz, 1843) (Figure 9.2F)

Terebridae
      Myurella taurina (Lightfoot, 1786)
      Myurellina floridana (Dall, 1889)
      Myurellina lindae (Petuch, 1987) (Figure 9.2C)
      Strioterebrum arcas (Abbott, 1954)

Turridae
      Polystira sunderlandi Petuch, 1987 (Figure 9.2L)

Drilliidae
      Fenimorea fucata (Reeve, 1845) (Figure 9.3M)
      Fenimorea kathyue Tippett, 1995 (north of the Dry Tortugas)
      Fenimorea petiti Tippett, 1995 (north of the Dry Tortugas)
      Fenimorea sunderlandi (Petuch, 1987) (Figure 9.2I)
      Neodrillia brunnescens (Rehder, 1943)
      Neodrillia woodringi (Bartsch, 1934)

2g. ECHINODERM FEEDERS (including echinoids [eccentric], asteroids, ophiuroids, and holothuroids)

Ranellidae
      Cymatium (Linatella) cingulatum (Lamarck, 1822)
      Cymatium (Septa) krebsii Mörch, 1877
      Cymatium (Septa) pharcidum Dall, 1889

Cassidae-Oocorythinae
      Eudolium thompsoni McGinty, 1955

Ficidae
      Ficus carolae Clench, 1945

3. SUSPENSION/FILTER FEEDERS

Gastropoda

Siliquariidae
      Siliquaria modesta Dall, 1889
      Siliquaria squamata Blainville, 1827

Turritellidae
      Torculoidella lindae Petuch, 1987 (Figure 9.1H)

Bivalvia

Arcidae
>   *Anadara baughmani* (Hertlein, 1951)

Glycymeridae
>   *Glycymeris americana* (DeFrance, 1826)
>   *Tucetona subtilis* Nicol, 1956

Mytilidae
>   *Dacrydium hendersoni* Salas and Gofas, 1997

Pectinidae
>   *Aequipecten lineolaris* (Lamarck, 1819)
>   *Euvola chazaliei* (Dautzenberg, 1900)
>   *Euvola raveneli* (Dall, 1898)
>   *Lindapecten lindae* Petuch, 1995 (Figure 9.2G, H)
>   *Spathochlamys benedicti* (Verrill and Bush, 1897)

Crassitellidae
>   *Crassinella dupliniana* (Dall, 1903)

Lucinidae
>   *Anodontia schrammi* (Crosse, 1876)
>   *Cavilinga blanda* (Dall, 1901)
>   *Ctena pectinella* (C.B. Adams, 1852)
>   *Lucinoma filosa* (Stimpson, 1851)
>   *Myrtea sagrinata* (Dall, 1886)
>   *Myrteopsis lens* (Verrill and Smith, 1880)
>   *Pleurolucina leucocyma* (Dall, 1886)

Chamidae
>   *Chama lactuca* Dall, 1886
>   *Chama sinuosa* Broderip, 1835

Cardiidae
>   *Acrosterigma magnum* (Linnaeus, 1758)
>   *Americardia lightbourni* Lee and Huber, 2012 (Figure 9.3B)
>   *Microcardium peramabile* (Dall, 1881)

Veneridae
>   *Chione cancellata* (Linnaeus, 1767) (deep-water analogue of the shallow-water
>   *C. elevata*)
>   *Lirophora clenchi* (Pulley, 1952)
>   *Lirophora paphia* (Linnaeus, 1758)

Figure 9.4 Mollusks of the *Rehderia schmitti* Assemblage. A = *Euvola marshallae* Petuch and Myers, new species; holotype, width 46 mm. Incorrectly referred to, by most workers, as *Euvola papyracea* (Gabb, 1873). That species is actually an early Miocene fossil species from the Baitoa Formation of the Dominican Republic, and this common deep-water scallop needed a new name (see the Systematic Appendix). B = *Cryptopecten phrygium* (Dall, 1886), length 32 mm. C = *Polystira albida* (Perry, 1811), length 70 mm. D = *Architectonica peracuta* (Dall, 1889), width 12 mm. E = *Siratus beauii* Fischer and Bernardi, 1857, form *branchi* Clench, 1953, length 72 mm. F = *Bartschia significans* Rehder, 1943, length 31 mm. G, H = *Rehderia georgiana* (Clench, 1946), length 80 mm. This volute ranges from North Carolina to the Miami Terrace off Key Largo. I = *Vexillum hendersoni* (Dall, 1927), length 13 mm. J = *Nodicostellaria laterculatum* (Sowerby II, 1874), length 11 mm. K = *Prunum frumari* Petuch and Sargent, 2011; holotype, length 12 mm. L = *Metulella columbellata* Dall, 1889, length 11 mm. M = *Murexiella hidalgoi* (Crosse, 1869), length 25 mm.

Lirophora varicosa (Sowerby II, 1853) (Figure 9.2K) (often incorrectly identified as
    Lirophora latilirata, which is actually a Pliocene fossil species from Virginia)
Pitarenus cordatus (Schwengel, 1951)

Tellinidae
Angulus agilis (Stimpson, 1857)
Angulus parameris (Boss, 1964)
Angulus probinus (Boss, 1964)
Cymatoica hendersoni Rehder, 1939
Elliptotellina americana (Dall, 1900)
Macoma pseudomera Dall and Simpson, 1901
Macoma tageliformis Dall, 1900
Merisca cristallina (Spengler, 1798)
Phyllodina squamifera (Deshayes, 1855)
Scissula consobrina (d'Orbigny, 1853)
Strigilla carnaria (Linnaeus, 1758)
Strigilla pisiformis (Linnaeus, 1758)

Semelidae
Abra americana Verrill and Bush, 1898
Abra lioica (Dall, 1881)

Pharidae
Ensis minor Dall, 1900

A review of the niche preferences of the members of the *Chicoreus rachelcarsonae*
Assemblage shows that the number of carnivores approximately equals the number of
suspension/filter feeders in the ecosystem.

Molluscan ecology of the deep softbottom terraces

The deep, narrow, shelf-like terraces that edge the seaward sides of the Florida Keys actually are a composite of three main geomorphological features: the Miami Terrace, the
Pourtales Terrace, and the Tortugas Valleys. The northernmost of these, the Miami Terrace,
extends from approximately the Dade-Broward County line southward to central Key
Largo. This deep linear feature is composed of eroded mid-Miocene limestone that has
undergone several erosional episodes during the late Miocene, Pliocene, and Pleistocene.
The upper section of the Miami Terrace (200- to 375-m depths) is heavily eroded, with
a karstic surface covered with sinkholes (Mullins and Neumann, 1979). In contrast, the
lower section (600- to 700-m depths) is composed of an eastward-dipping slope and is
covered with thick layers of pelagic sediments. The central Pourtales Terrace, the largest
of the deep features with a length of 213 km, extends from southern Key Largo westward
to the Marquesas Keys and ranges in depth from 200 to 450 m. Like the Miami Terrace
to the north, the Pourtales Terrace is heavily eroded, with a karstic surface and extensive
systems of deep sinkholes (Reed et al., 2005). The southernmost feature, the complex of the
Tortugas Valleys, extends southward of the Dry Tortugas islands at depths of 200–300 m.
These deep, parallel depressed features represent a series of drowned river channels and
valleys that formed from drainage off the emergent Florida Peninsula during Pleistocene
extreme sea-level drops.

    Because of the great depths and the expense and technical difficulty of collecting
specimens, the malacofauna of the three deep terraces is still poorly known. Exploratory
work involving dredging from research vessels and direct sampling from submersible

vehicles (Reed et al., 2005), however, has shown that the invertebrate faunas of these areas are extremely rich, with many undescribed species. At depths of 250–300 m, the terrace seafloors are covered with dense growths of large sponges, small stalked crinoids, and stylasterine hydrocorals, and these form the basis of an ecologically diverse and species-rich ecosystem. Although poorly sampled and still virtually unknown, the molluscan fauna of these Florida Keys deep-water terrace ecosystems has been found to contain a large number of endemic species. Primary among these Keys-restricted taxa are gastropods such as the volute *Rehderia schmitti* (Figure 9.5C, D; namesake of the entire deep-water malacofauna, the *Rehderia schmitti* Assemblage); the pisaniiform buccinid *Bartschia frumari* (Figure 9.4F); the marginellids *Prunum redfieldi* and *Prunum frumari* (Figure 9.4K); the cystiscids *Canalispira kerni* and *Cysticus microgonia*; the olivellid *Niteoliva moorci*; and the olivids *Americoliva recourti* new species (Figure 9.6A–C) and *Americoliva matchetti* new species (Figure 9.6D–F) (for the descriptions of the new olives, see the Systematic Appendix). The malacofauna of the deep terraces, even though only partially known, has been found to be one of the richest in the Florida Keys area, with over 165 species of gastropods and over 75 bivalves. An excellent overview and iconography of the deep-water cone shells (Conidae and Conilithidae) of the Florida Keys and adjacent areas was given by Tucker (2013).

Thick accumulations of carbonate mud and pelagic sediments characteristically blanket the planar sections of the deep terraces, and these create the perfect environment for a rich bivalve fauna. On some of the wider terraces near the Marquesas Keys and Dry Tortugas, large aggregations of scallops (Pectinidae and Propeamussiidae) often carpet the seafloor. Shoals of these mobile swimmers are continuously moving from place to place, generally in response to seasonal changes in plankton food resources caused by ephemeral upwelling systems. Four species of large pectinids dominate the muddy seafloor: *Aequipecten glyptus, Euvola chazaliei, Euvola laurenti,* and *Euvola marshallae* new species (Figure 9.4A; see Systematic Appendix). The two last-mentioned scallops range throughout the Gulf of Mexico and Caribbean Sea and are particularly abundant north of the Dry Tortugas, along western Florida, and off the Yucatan Peninsula in depths of around 200–250 m. The *Euvola* and *Aequipecten* pectinids occur with a large fauna of small glass scallops of the family Propeamussiidae comprising six species in five different genera (listed in the section on the biodiversity of the *Rehderia schmitti* Assemblage).

The deep terrace muddy seafloors also house an unusually large fauna of primitive palaeotaxodont and cryptodont bivalves, comprising four families (Nuculidae, Solemyidae, Nuculanidae, and Yoldiidae), seven genera, and at least 12 species. Some of the more abundant and conspicuous of these include the nuculids *Nucula crenulata* and *Nucula delphinodonta*; the nuculanids *Propeleda carpenteri, Ledella sublaevis,* and *Nuculana acuta*; and the yoldiid *Yoldia liorhina*. These primitive relictual forms occur with an extremely rich fauna of classic deep-water bivalves, including the arcid *Bathyarca glomerula*; the limopsids *Limopsis cristata, Limopsis minuta,* and *Limopsis sulcata*; the limids *Divarilima albicoma, Limea bronniana,* and *Limatula setifera*; the astartids *Astarte nana, Astarte smithii,* and *Astarte subaequilatera*; the pandorids *Pandora inflata* and *Pandora bushiana*; the verticordiids *Spinosipella acuticostata* and *Euciroa elegantissima*; and the cuspidariids *Cuspidaria rostrata, Cardiomya perrostrata, Myonera gigantea,* and *Plectodon granulatus*. The deep-water scallops and these shallowly infauna bathyal bivalves serve as the primary food source for a large complement of molluscivorous predatory gastropods, including drillers such as the naticids *Neverita nubila, Lunatia fringilla, Natica perlineata,* and *Sinum minor* (Figure 9.7L), and the muricids *Siratus beauii* (Figure 9.4E), *Siratus articulatus* (Figure 9.7J), *Murexiella hidalgoi* (Figure 9.4M), *Vokesimurex cabritii* (Figure 9.7E), and *Poirieria nuttingi*; and general molluscivores such as the fasciolariids *Harasewychia benthalis, Harasewychia aepynotus,* and *Fusinus halistreptus*; the

buccinids *Liomesus stimpsoni, Ptychosalpinx globulus,* and *Bartschia significans* (Figure 9.4F); the acteonids *Acteon finlayi, Acteon danaida, Ovulacteon meekii,* and *Bullina exquisita;* and the large cylichnids *Scaphander punctostriatus* and *Scaphander (Sabatia) bathymophila.*

A large fauna of carnivorous volutoidean gastropods, comprising the families Marginellidae, Volutidae, Volutomitridae, Mitridae, Costellariidae, Cancellariidae, Olivellidae, and Olividae, is also present on the muddy terraces and competes with the naticids, muricids, and buccinids for molluscan prey. Of these, the scaphelline volutes are the most conspicuous and most frequently encountered, often occurring in such abundance that their dead shells make up a pavement across the seafloor. Although often incorrectly placed in the genus *Scaphella* by many workers, the seven species of volutes found on the Keys terraces actually belong to three distinct genera: *Aurinia* (large inflated shells with twisted, curled calcarellas that have sharply pointed tips); *Rehderia* (medium-size, slender species with large, rounded, dome-like calcarellas); and *Clenchina* (small shells, often with stepped and beaded spire whorls, and with small, rounded, protracting calcarellas). The genus *Scaphella,* as typified by *Scaphella junonia* and its subspecies (*elizabethae, johnstoneae,* and *butleri*), have proportionally small, rounded calcarellas and heavily ribbed early whorls. These larger, heavier, and more colorful volutes live in much shallower water than do *Clenchina, Rehderia,* and *Aurinia,* often occurring as shallow as 3 m deep (as in the case of *Scaphella junonia elizabethae;* see Chapter 6). Of the seven deep-water volutes, only *Rehderia schmitti* (Figure 9.5C, D; the namesake of the molluscan assemblage) and *Clenchina dohrni* (Figure 9.5G, H) are endemic to the Florida Keys area, most commonly collected off the Dry Tortugas (the type locality for *schmitti*) and Key West. The largest of the deep terrace scaphelline volutes, *Aurinia dubia* (Figure 9.5A, B), ranges throughout the Gulf of Mexico, and its presence off the Florida Keys marks its far southeastern-most occurrence. Two other species of the genus *Clenchina, C. marionae* (Figure 9.5E, F) and *C. florida* (9.5I, J), both have relatively wide ranges, extending from southeastern Florida to the northern Gulf of Mexico and also to northern Cuba. The beautifully banded *Clenchina gouldiana* (Figure 9.5K, L) has the widest range of all the deep-water volutes, extending from North Carolina to the northern Gulf of Mexico. The heavily ribbed *Rehderia georgiana* (Figure 9.4G, H; normally found off the Carolinas and Georgia) ranges only to off the Upper Keys, and this area marks the southernmost extension of its distribution.

The smaller volutoideans of the *Rehderia schmitti* Assemblage, although often not as numerous as the scaphelline volutes, comprise a species-rich complex of 16 genera. The largest of these volutoidean families is the Marginellidae, which is represented by such distinctive species as *Prunum redfieldi* and *Prunum frumari* (Figure 9.4K), both of which are endemic to the Keys, and the bizarre, twisted *Prunum torticula.* Another large family, the Olivellidae, contains species that often occur together in large aggregations on the muddy seafloors and includes such characteristic taxa as *Jaspidella miris, Macgintiella watermani, Niteoliva moorei* (endemic to the Florida keys), and *Callianax thompsoni.* These abundant

Figure 9.5 Volutes of the *Rehderia schmitti* Assemblage. A, B = *Aurinia dubia* (Broderip, 1827), length 135 mm. The closely related but much larger *Aurinia kieneri* (Clench, 1946) is confined to the Gulf of Mexico, from western Florida to Texas (see Petuch, 2013: 67). C, D = *Rehderia schmitti* (Bartsch, 1931), length 82 mm. Note the large, rounded, knoblike protoconch and calcarella on the genus *Rehderia.* This character separates *Rehderia* from the genus *Aurinia,* which has a twisted, curled calcarella. Endemic to the Florida Keys. E, F = *Clenchina marionae* (Pilsbry and Olsson, 1953), length 33 mm. G, H = *Clenchina dohrni* (Sowerby III, 1903), length 52 mm. Endemic to the Florida Keys. I, J = *Clenchina florida* (Clench and Aguayo, 1940), length 57 mm. K, L = *Clenchina gouldiana* (Dall, 1887), length 54 mm.

small olivellids occur sympatrically with two newly discovered olivids, *Americoliva recourti* new species (Figure 9.6A–C) and *Americoliva matchetti* new species (Figure 9.6D–F) (see the Systematic Appendix for the descriptions of these new species), with which they probably compete for small crustacean and molluscan prey and carrion. Of the two new deepwater olive shells, *Americoliva recourti* has the widest distribution, ranging from southern Palm Beach County southward to at least the Middle Keys. The smaller and stumpier *Americoliva matchetti*, which somewhat resembles a tiny version of the Bahamian and West Indian *Americoliva reticularis* (see Petuch, 2013: 21), has a more limited distribution, only ranging from Broward County to Key Largo. The olives and olivellids occur together with other classic deep-water volutoideans, such as the volutomitrids *Microvoluta blakeana* and *Microvoluta laevior*, and costellariids such as *Vexillum hendersoni* (Figure 9.4I), *Vexillum styria*, and *Nodicostellaria laterculata* (Figure 9.4J).

The muddy sediments of the Keys deep terraces are rich in organic detritus that has trickled downslope from the living reefs high above. This potential food resource has been utilized by a large fauna of gastropod detritivores and bacteria feeders such as the solariellids *Solariella lacunella*, *Solariella pourtalesi*, and *Suavotrochus lubrica*; the margaritids *Gaza superba* and *Microgaza rotella*; the architectonicids *Architectonica sunderlandi* and *Architectonica peracuta* (Figure 9.4D); and the deep-water xenophorids *Trochotugurium longleyi* and *Tugurium caribaeum*. Larger pieces of organic detritus and carrion are the principal food resource for a suite of small deep-water scavengers, including the nassariid mud snails *Uzita hotessieri* and *Uzita scissurata* and the columbellids *Metulella columbellata* (Figure 9.4L) and *Nassarina bushiae*. On the Pourtales and Miami Terraces, sponge beds that are adjacent to the mud seafloors and that line the walls of deep sinkholes (Reed et al., 2005) form the substrate and food resource for an interesting fauna of spongivorous gastropods. Some of these include the pleurotomariids *Entemnotrochus adansonianus*, *Perotrochus amabilis*, and *Bayerotrochus midas* and the large calliostomatids *Calliostoma (Kombologion) psyche*, *Calliostoma (Kombologion) bairdii*, and *Calliostoma (Kombologion) oregon*.

The soft substrates of the deep terraces also harbor a large fauna of polychaete and sipunculid worms, as evidenced by the extremely large resident fauna of vermivorous gastropods that prey on them. On the Tortugas Valleys and Pourtales Terrace alone, over 48 vermivorous gastropods are known to occur, and these include species belonging to 10 separate families: the Conidae, Conilithidae, Terebridae, Cochlespiridae, Turridae,

Figure 9.6 Olivid and Conilithid Gastropods from the *Rehderia schmitti* Assemblage. A, B = *Americoliva recourti* Petuch and Myers, new species; holotype, length 23.5 mm (see Systematic Appendix for the complete description). C = *Americoliva recourti* Petuch and Myers, new species. Paratype, length 22 mm. D, E = *Americoliva matchetti* Petuch and Myers, new species; holotype, length 18.5 mm (see Systematic Appendix for the complete description). F = *Americoliva matchetti* Petuch and Myers, new species. Paratype, length 15 mm. G, H = *Kohniconus* new species (?), length 16.5 mm. This unusual deep-water cone, from 420-m depth along the edge of the Pourtales Terrace, resembles a juvenile of the shallower-water *K. delessertii*. Although similar in general appearance, this possible new species differs in having a more pronounced pyriform shape, in having stronger coronations on the shoulder carina, in having stronger and more prominent spiral cords around the anterior end, and in being pure white in color, with 12 rows of faint, very tiny tan dots on the body whorl. Although only known from this single specimen, a sampling of more and better material may eventually demonstrate that this shiny white, very-deep-water cone represents an unnamed endemic *Kohniconus* species (collected by deep-water "Royal Red Shrimp" fishermen off Vaca Key, Central Florida Keys).

Figure 9.7 Gastropods of the *Rehderia schmitti* Assemblage. A = *Conasprelloides stimpsoni* (Dall, 1902), length 51 mm. B = *Cochlespira elegans* (Dall, 1881), length 35 mm. C = *Gemmula periscelida* (Dall, 1889), length 43 mm. D = *Conasprelloides villepinii* (Fischer and Bernardi, 1857), length 35 mm. E = *Vokesimurex cabritii* (Bernardi, 1859), length 62 mm. F = *Strioterebrum rushii* (Dall, 1889), length 15 mm. G = *Polystira florencae* Bartsch, 1934, length 22 mm. H = *Strioterebrum lutescens* (Dall, 1889), length 14 mm. I = *Strioterebrum limatulum* (Dall, 1889), length 14 mm. J = *Siratus articulatus* (Reeve, 1845), length 60 mm. K = *Ficus carolae* Clench, 1845, length 66 mm. L = *Sinum minor* (Dall, 1889), length 9 mm.

Clathurellidae, Drilliidae, Pseudomelatomidae, Crassispiridae, and Raphitomidae. Some of the largest worm-eating species are found in the families Conidae and Conilithidae, and these include such prominent taxa as *Conasprelloides stimpsoni* (Figure 9.7A), *Conasprelloides villepini* (Figure 9.7D), and *Conasprelloides cancellatus* (all Conidae) and *Kohniconus* new species (?) (Figure 9.6G, H), *Dalliconus rainesae*, and *Dalliconus mcgintyi* (all Conilithidae) (see Tucker, 2013). These cone shells live together with several poorly known and rarely seen deep-water terebrids, including *Strioterebrum rushii* (Figure 9.7F), *Strioterebrum lutescens* (Figure 9.7H), *Strioterebrum limatulum* (Figure 9.7I), and *Fusoterebra benthalis*. An exceptionally large fauna of turroidean gastropods also occurs with the cone shells and terebrids and typically includes species such as the turrids *Polystira albida* (Figure 9.4C), *Polystira florencae* (Figure 9.7G), and *Gemmula periscelida* (Figure 9.7C); the cochlespirids *Cochlespira elegans* (Figure 9.7B), *Cochlespira radiata*, *Leucosyrinx tenoceras*, and *Leucosyrinx subgrundifera*; the clathurellids *Bathytoma viabrunnea*, *Glyphostoma gabbi*, and *Glyphostoma pilsbryi*; the zonulispirids *Compsodrillia halistrephis* and *Compsodrillia eucosmia*; the crassipirid *Hindsiclava alesidota*; and the mangeliids *Mangelia exsculpta*, *Stellatoma antonia*, *Gymnobela blakeana*, and *Stellatoma monocingulata*.

A large echinoderm fauna is also present on the muddy seafloors of the deep terraces, including echinoids, holothroids, ophiuroids, crinoids (both stalked and free-living comatulids), and asteroids. These serve as the main prey items of an interesting fauna of tonnoidean gastropods, including the cassids *Sconsia striata* and *Echinophoria coronadoi* (both feeding on echinoids); the oocorythines *Eudolium crosseanum*, *Oocorys bartschi*, and *Benthodolium abyssorum*; and the ficid *Ficus carolae* (Figure 9.7K). This distinctive echinoderm-feeding faunule occurs together with a number of deep-water cnidarian-feeding gastropods, some of which include the large and heavily fluted epitoniid *Stenorhytis pernobilis* and the smaller epitoniids *Opaliopsis nitida* and *Opaliopsis opalina*. All of these are probably feeding suctorially on large sand-dwelling zoantherian sea anemones.

Biodiversity of the Rehderia schmitti Assemblage

The members of the deep terrace-dwelling *Rehderia schmitti* Assemblage are listed here by feeding type and ecological niche preferences. As in the previous section of this chapter, this list must be considered only preliminary because future deep-water dredging will undoubtedly uncover many more new and unusual taxa. For illustrations of the species not shown here, see the iconographies of Abbott (1974; for gastropods) and Mikkelsen and Bieler (2008; for bivalves). Because this ecosystem exists at depths below the photic zone, no plants or algae are present, and herbivorous gastropods cannot live here. This niche is filled by detritivores, which feed on plant and animal matter that tumbles down the deep reef talus slope and accumulates on the muddy planar surface of the terraces.

1. **HERBIVORES/DETRITIVORES (including particulate organic feeders and bacterial film grazers)**
   Gastropoda
   Fissurellidae
   Emarginula sicula Gray, 1825
   Emarginula tuberculosa Libassi, 1859
   Solariellidae
   Solariella aegleis Watson, 1879
   Solariella lacunella (Dall, 1881)

Solariella lamellosa Verrill and Smith, 1880
Solariella lissocoma (Dall, 1881)
Solariella periscopia (Dall, 1927)
Solariella pourtalesi Clench and Aguayo, 1939
Suavotrochus lubricus (Dall, 1881)

Margaritidae
Gaza superba (Dall, 1881)
Microgaza inornata Dall, 1881
Microgaza rotella Dall, 1881

Architectonicidae
Architectonica peracuta (Dall, 1889) (Figure 9.4D)
Architectonica sunderlandi Petuch, 1987
Solatisonax borealis (Verrill and Smith, 1880) (may feed on cnidarians)

Xenophoridae
Trochotugurium longleyi (Bartsch, 1931)
Tugurium caribaeum (Petit, 1856)

2. CARNIVORES
Gastropoda
2a. SPONGIVORES
Pleurotomariidae
Bayerotrochus midas (Bayer, 1965)
Entemnotrochus adansonianus (Crosse and Fischer, 1861)
Perotrochus amabilis (Bayer, 1963)

Haliotidae
Haliotis pourtalesi Dall, 1881

Calliostomatidae
Calliostoma (Kombologion) bairdii Verrill and Smith, 1880
Calliostoma (Kombologion) oregon Clench and Turner, 1960
Calliostoma (Kombologion) psyche Dall, 1889

2b. GENERAL CARNIVORES/SCAVENGERS
Personidae (feeding preferences problematical)
Distorsio perdistorta Fulton, 1938

Buccinidae
Antillophos virginiae (Schwengel, 1942)
Agassitula agassizi (Clench and Aguayo, 1941)
Bartschia significans Rehder, 1943 (Figure 9.4F) (*B. Frumari* Garcia is a synonym)
Belomitra pourtalesi (Dall, 1881) (often collected below 500-m depth)
Colus rushii (Dall, 1889)
Eosipho canetae (Clench and Aguayo, 1941)
Liomesus stimpsoni Dall, 1889
Manaria burkeae Garcia, 2008
Ptychosalpinx globulus (Dall, 1889)
Retimohnia carolinensis (Verrill, 1884)

Columbellidae
Cosmioconcha rikae Monsecour and Monsecour, 2006 **D**
Metulella columbellata (Dall, 1889) (Figure 9.4L)
Nassarina bushiae (Dall, 1889)

Nassariidae
> *Uzita hotessieri* (d'Orbigny, 1845)
> *Uzita scissurata* (Dall, 1889)

Olivellidae
> *Calliunax thompsoni* (Olsson, 1956)
> *Jaspidella miris* Olsson, 1956
> *Macgintiella rotunda* (Dall, 1889)
> *Macgintiella watermani* (McGinty, 1940)
> *Niteoliva moorei* (Abbott, 1951)
> *Olivella mcgintyi* Olsson, 1956
> *Olivina bullula* (Reeve, 1850)

Olividae
> *Americoliva matchetti* Petuch and Myers, new species (Figure 9.6D–F) (see Systematic
> Appendix)
> *Americoliva recourti* Petuch and Myers, new species (Figure 9.6A–C) (see Systematic
> Appendix)

Volutidae-Scaphellinae
> *Aurinia dubia* (Broderip, 1827) (Figure 9.5A, B)
> *Clenchina dohrni* (Sowerby III, 1903) (Figure 9.5G, H)
> *Clenchina florida* (Clench and Aguayo, 1940) (Figure 9.5I, J)
> *Clenchina gouldiana* (Dall, 1887) (Figure 9.5K, L)
> *Clenchina marionae* (Pilsbry and Olsson, 1953) (Figure 9.5E, F)
> *Rehderia georgiana* (Clench, 1946) (Figure 9.4G, H; Cape Hatteras to Key Largo)
> *Rehderia schmitti* (Bartsch, 1931) (Figure 9.5C, D)

Volutomitridae
> *Microvoluta blakeana* (Dall, 1889)
> *Microvoluta laevior* (Dall, 1889)

Marginellidae
> *Dentimargo yucatecanus* (Dall, 1881)
> *Eratoidea watsoni* (Dall, 1881)
> *Hyalina pallida* (Linnaeus, 1758)
> *Prunum cassis* (Dall, 1889)
> *Prunum frumari* Petuch and Sargent, 2011 (Figure 9.4K)
> *Prunum redfieldi* (Tryon, 1882)
> *Prunum torticulum* (Dall, 1881)
> *Prunum virginianum* (Conrad, 1868)
> *Volvarina styria* Dall, 1889
> *Volvarina subtriplicata* (d'Orbigny, 1842)

Cystiscidae (Persiculinae and Cystiscinae)
> *Canalispira kerni* Garcia, 2007
> *Cysticus microgonia* (Dall, 1927)

Acteonidae
> *Acteon cumingii* A. Adams, 1854
> *Acteon danaida* Dall, 1881
> *Acteon delicatus* Dall, 1889
> *Acteon finlayi* McGinty, 1955
> *Acteon incisus* Dall, 1881
> *Acteon melampoides* Dall, 1881

Acteon perforatus Dall, 1881
Acteon pusillus (Forbes, 1844)
Ovulacteon meekii Dall, 1889
Bullinidae
Bullina exquisita McGinty, 1955
Cylichnidae
Scaphander punctostriatus Mighels, 1841
Scaphander watsoni Dall, 1881
Scaphander (Sabatia) bathymophila Dall, 1881
Acteocinidae
Acteocina bullata (Kiener, 1834)
Bullidae
Bulla eburnea Dall, 1881
2c. **MOLLUSCIVORES (including drilling predators)**
Naticidae
Lunatia fringilla (Dall, 1881)
Lunatia perla (Dall, 1881)
Natica castrensis Dall, 1889
Natica perlineata Dall, 1889
Neverita nubila (Dall, 1925)
Sigatica carolinensis (Dall, 1889)
Sinum minor (Dall, 1889) (Figure 9.7L)
Muricidae (Muricinae, Ergalitaxinae, Muricopsinae, Ocenebrinae)
Murexiella hidalgoi (Crosse, 1869) (Figure 9.4M)
Orania grayi (Dall, 1889)
Paziella pazi (Crosse, 1869)
Pazinotus stimpsoni (Dall, 1889)
Poirieria nuttingi (Dall, 1896)
Pterynotus bushae E. Vokes, 1970
Pterynotus tristichus (Dall, 1889)
Siratus articulatus (Reeve, 1845) (Figure 9.7J)
Siratus beauii (Fischer and Bernardi, 1857) (Figure 9.4E)
Trossulasalpinx macra (Verrill, 1887) (the genus was described from fossil taxa)
Vokesimurex cabritii (Bernardi, 1859) (Figure 9.7E)
Vokesimurex tryoni (Hidalgo, 1880)
Fasciolariidae (Subfamily Fusininae)
Fusinus excavatus (Sowerby II, 1880) (= *eucosmius* Dall)
Fusinus halistreptus (Dall, 1889)
Fusinus schrammi (Crosse, 1865)
Harasewychia aepynota (Dall, 1889)
Harasewychia alcimus (Dall, 1889)
Harasewychia amianta (Dall, 1889)
Harasewychia amphiurgus (Dall, 1889)
Harasewychia benthalis (Dall, 1889)
Harasewychia rushii (Dall, 1889)
Costellariidae
Nodicostellaria laterculata (Sowerby II, 1874) (Figure 9.4J)
Vexillum hendersoni (Dall, 1927) (Figure 9.4I)

    Vexillum styliolum (Dall, 1927)

    Vexillum styria Dall, 1889

    Vexillum wandoense (Holmes, 1860)

2d. SUCTORIAL FEEDERS (problematical; on benthic elasmobranchs?)

Cancellariidae

    Axelella agassizii (Dall, 1889)

2e. ECHINODERM FEEDERS (including echinoids [eccentric], holothurians, ophiuroids, and asteroids)

Cassidae-Cassinae

    Echinophoria coronadoi (Crosse, 1867)

    Sconsia striata (Lamarck, 1816) (incorrectly called *S. grayi* Adams, 1855, by some workers)

Cassidae-Oocorythinae

    Benthodolium abyssorum (Verrill and Smith, 1884)

    Eudolium crosseanum (Monterosato, 1869)

    Oocorys bartschi Rehder, 1943

    Oocorys sulcata Fischer, 1883

Ficidae

    Ficus carolae Clench, 1945) (Figure 9.7K)

2f. CNIDARIAN FEEDERS (including suctorial forms)

Ovulidae-Simniinae (feeding on antipatharian and spiropatharian octocorals)

    Calcarovula piragua (Dall, 1889)

    Pseudocyphoma aureocincta (Dall, 1889)

Epitoniidae

    Amaea mitchelli (Dall, 1896) **D**

    Amaea retifera (Dall, 1889)

    Asperiscala babylonium (Dall, 1889)

    Asperiscala frielei (Dall, 1889)

    Asperiscala fractum (Dall, 1927)

    Asperiscala polacia (Dall, 1889)

    Asperiscala pourtalesi (Verrill and Bush, 1880)

    Asperiscala rushii (Dall, 1889)

    Boreoscala blainei Clench and Turner, 1953

    Cylindriscala andrewsi (Verrill, 1882)

    Cylindriscala watsoni (deBoury, 1911)

    Depressiscula nitidella (Dall, 1889)

    Foratiscala formossisima (Jeffreys, 1884)

    Nodiscala eolis Clench and Turner, 1950

    Opalia abbotti Clench and Turner, 1952

    Opaliopsis atlantis Clench and Turner, 1952

    Opaliopsis cania (Dall, 1927)

    Opaliopsis concava (Dall, 1889) **E**

    Opaliopsis nitida (Verrill and Smith, 1885)

    Opaliopsis opalina (Dall, 1927)

    Stenorhytis pernobilis (Fischer and Bernardi, 1857)

Muricidae-Magilinae

    Babelomurex fax (Bayer, 1971) (possibly feeding on zoantherians)

2g. VERMIVORES (including feeders on polychaetes, sipunculids, and nemerteans)
Turbinellidae-Columbariinae
Peristarium aurora (Bayer, 1971)
Peristarium electra (Bayer, 1971)
Peristarium merope (Bayer, 1971)
Mitridae
Mitra (Cancilla) antillensis Dall, 1889
Subcancilla straminea (A. Adams, 1853)
Conidae
Conasprelloides cancellatus (Hwass, 1792)
Conasprelloides stimpsoni (Dall, 1902) (Figure 9.7A) (*C. levistimpsoni* Tucker is a synonym)
Conasprelloides villepini (Fischer and Bernardi, 1857) (Figure 9.7D)
Conilithidae
Dalliconus mcgintyi (Pilsbry, 1955)
Dalliconus rainesae (McGinty, 1953)
Kohniconus new species? (Figure 9.6G, H)
Terebridae
Fusoterebra benthalis (Dall, 1899)
Myurellina floridana (Dall, 1889)
Strioterebrum limatulum (Dall, 1889) (Figure 9.7I)
Strioterebrum lutescens (Dall, 1889) (Figure 9.7H)
Strioterebrum rushii (Dall, 1889) (Figure 9.7F)
Cochlespiridae
Cochlespira elegans (Dall, 1881) (Figure 9.7B)
Cochlespira radiata (Dall, 1889)
Leucosyrinx subgrundifera (Dall, 1888)
Leucosyrinx tenoceras Dall, 1889
Leucosyrinx verrillii (Dall, 1881)
Turridae
Gemmula periscelida (Dall, 1889) (Figure 9.7C)
Polystira albida (Perry, 1811) (Figure 9.4C)
Polystira florencae Bartsch, 1934 (Figure 9.7G)
Polystira tellea (Dall, 1889)
Clathurellidae
Bathytoma viabrunnea (Dall, 1889)
Drilliola loprestiana (Calcara, 1841) (= *Microdrillia comatotropis*)
Glyphostoma dentifera Gabb, 1872
Glyphostoma gabbi Dall, 1889
Glyphostoma gratula (Dall, 1881)
Glyphostoma pilsbryi (Schwengel, 1940)
Lioglyphostoma oenoa Bartsch, 1934
Miraclathurella herminea Bartsch, 1934
Mitrolumna biplicata Dall, 1889
Mitrolumna haycocki (Dall and Bartsch, 1911)
Tennaturris inepta (E.A. Smith, 1882)
Drilliidae
Drillia acrybia (Dall, 1889)
Drillia pharcida (Dall, 1889)

Crassispiridae
 Hindsiclava alesidota (Dall, 1889)
Zonulispiridae
 Compsodrillia acestra (Dall, 1889)
 Compsodrillia disticha Bartsch, 1934
 Compsodrillia eucosmia (Dall, 1889)
 Compsodrillia haliostrephis (Dall, 1889)
Raphitomidae
 Anticlinura toreumata (Dall, 1889)
 Daphnella eugrammata (Dall, 1902)
 Daphnella margaretae Lyons, 1972
 Daphnella morra (Dall, 1881)
 Eubela mcgintyi Schwengel, 1943
 Gymnobela agassizii (Verrill and Smith, 1880)
 Gymnobela blakeana (Dall, 1889)
 Mangelia bartletti (Dall, 1889)
 Mangelia exsculpta (Watson, 1881)
 Mangelia pelagia (Dall, 1881)
 Mangelia pourtalesi (Dall, 1881)
 Mangelia scipio Dall, 1889
 Mangelia subsida (Dall, 1881)
 Stellatoma antonia (Dall, 1881)
 Stellatoma monocingulata (Dall, 1889)

3. **SUSPENSION/FILTER FEEDERS**
 Gastropoda
 Capulidae
  Capulus ungaricus (Linnaeus, 1767)
 Hipponicidae
  Malluvium benthophilum (Dall, 1889) (lives attached to the spines of cidaroid echinoids)
 Bivalvia
 Nuculidae
  Eunucula aegeensis (Forbes, 1844)
  Nucula crenulata A. Adams, 1856
  Nucula delphinodonta Mighels and C.B. Adams, 1842
 Nuculanidae (feeding on surface films by using specialized palps)
  Ledella sublaevis Verrill and Bush, 1898
  Nuculana acuta (Conrad, 1832)
  Nuculana jamaicensis (d'Orbigny, 1853)
  Nuculana semen (E.A. Smith, 1885)
  Nuculana verrilliana (Dall, 1886)
  Nuculana vitrea (d'Orbigny, 1853)
 Arcidae
  Bathyarca glomerula (Dall, 1881)
  Bathyarca sagrinata (Dall, 1886)
 Limopsidae
  Limopsis aurita (Brocchi, 1814)
  Limopsis cristata Jeffreys, 1876
  Limopsis minuta (Philippi, 1836)
  Limopsis sulcata Verrill and Bush, 1898

Limidae
Divarilima albicoma (Dall, 1886)
Limea bronniana Dall, 1886
Limatula confusa (E.A. Smith, 1885)
Limatula setifera Dall, 1886
Limatula subovata (Jeffreys, 1876)
Pectinidae
Aequipecten glyptus (Verrill, 1882)
Cryptopecten phrygium (Dall, 1886) (Figure 9.4B)
Euvola chazaliei (Dautzenberg, 1900)
Euvola laurenti (Gmelin, 1791)
Euvola marshallae Petuch and Myers, new species (Figure 9.4A) [incorrectly referred to as *Euvola papyracea* (Gabb, 1873), which is actually an early Miocene fossil from the Baitoa Formation of the Dominican Republic]
Propeamussiidae
Cyclopecten thalassinus (Dall, 1886)
Hyalopecten strigillatus (Dall, 1889)
Parvamussium pourtalesianum (Dall, 1886)
Parvamussium sayanum (Dall, 1886)
Propeamussium cancellatum (E.A. Smith, 1885)
Similipecten nanus (Verrill and Bush, 1897)
Astartidae
Astarte nana E.A. Smith, 1881
Astarte smithii Dall, 1866
Astarte subaequilatera Sowerby II, 1854
Pandoridae
Pandora bushiana Dall, 1886
Pandora glacialis Leach, 1819
Pandora inflata Boss and Merrill, 1965
Periplomatidae
Periploma tenerum (P. Fischer, 1882)
Verticordiidae
Euciroa elegantissima (Dall, 1881)
Haliris fischeriana (Dall, 1881)
Spinosipella acuticostata (Philippi, 1844)
Trigonulina ornata d'Orbigny, 1853
Poromyidae
Poromya albida Dall, 1886
Poromya granulata (Nyst and Westendorp, 1839)
Poromya rostrata Rehder, 1943
Cuspidariidae
Cardiomya alternata (d'Orbigny, 1853)
Cardiomya costellata (Deshayes, 1833)
Cardiomya costellata (Deshayes, 1833)
Cardiomya gemma Verrill and Bush, 1898
Cardiomya ornatissima (d'Orbigny, 1853)
Cardiomya perrostrata (Dall, 1881)

> *Cardiomya striata* (Jeffreys, 1876)
> *Cuspidaria obesa* (Loven, 1846)
> *Cuspidaria rostrata* (Spengler, 1793)
> *Myonera gigantea* (Verrill, 1884)
> *Myonera lamellifera* (Dall, 1881)
> *Myonera paucistriata* Dall, 1886
> *Plectodon granulatus* (Dall, 1881)

Lucinidae
> *Pleurolucina leucocyma* (Dall, 1886)
> *Pleurolucina sombrerensis* (Dall, 1886)

Thyasiridae (possibly having symbiotic chemotrophic bacteria)
> *Thyasira grandis* (Verrill and Smith, 1885)
> *Thyasira trisinuata* (d'Orbigny, 1853)

Lasaeidae
> *Erycina periscopiana* Dall, 1899

Sportellidae
> *Basterotia elliptica* (Recluz, 1850)

4. **CHEMOSYNTHETIC FEEDERS** (containing symbiotic chemoautotrophic bacteria)
Bivalvia
Solemyidae
> *Solemya occidentalis* Deshayes, 1857

A review of the niche preferences of the members of the *Rehderia schmitti* Assemblage shows that the carnivorous gastropods completely dominate the ecosystem by over a two-to-one ratio.

Mollusks of the open oceanic pleuston

The southern tip of the Floridian Peninsula extends into some of the most dynamic oceanographic conditions found anywhere in the tropical western Atlantic. This unique open oceanic world is the result of the confluence of three separate currents: the Gulf Loop Current, flowing southward along the southwestern coast of Florida and then across the Dry Tortugas (and producing the localized Tortugas Gyre); the Florida Current, flowing eastward over the Florida Keys and through the Florida Strait between Cuba and the Florida Keys; and the Caribbean Current, flowing northward out of the Caribbean Basin and entering the Florida Strait near Cay Sal Bank, Bahamas. These three currents merge north of Cay Sal to produce the Gulf Stream, which flows northward out of the Florida Strait and roughly parallels the coastline of the southeastern United States. Off Cape Hatteras, North Carolina, the Gulf Stream makes a sharply angled turn to the northeast and flows outward into the open Atlantic Ocean.

During its flow through the northern section of the Florida Strait, between the Floridian Peninsula and the Bahamas Banks, the Gulf Stream comes closest to land along a stretch of coast extending from northern Key Largo to Jupiter, Palm Beach County. Because of the proximity of this swiftly flowing current to the coastline, the shorelines of the northern Keys and Dade, Broward, and Palm Beach Counties are frequently inundated with immense masses of open oceanic pleuston. These organisms, which float directly on the sea surface or hang from the surface tension, exist as vast flotillas out in the open Gulf Stream waters but are blown up onto the beaches during storms or by strong easterly winds (Figure 9.9). Typically, the pleuston is dominated by floating siphonophore hydrozoans such as *Physalia physalis* (Portuguese Man-of-War), *Velella velella* (By-the-Wind Sailor), and *Porpita porpita* (Blue Button). These bright blue and purple pleustonic colonial hydrozoans, which hang from the surface by either gas-filled floats (like *Physalia*) or by thin lightweight disks that cling to the surface tension (like *Velella* and *Porpita*), often form immense densely packed flotillas that are referred to by local marine biologists as the "Great Blue Fleet." These hydrozoan aggregations serve as a food resource for a small fauna of pleustonic shelled macrogastropods, all in the family Janthinidae (the "Violet Snails"). The bright purple and pink janthinids receive their shell color from the pigments of their blue siphonophore prey and float, upside-down, from bubble rafts made of stiff mucus.

The five species of janthinids that occur off the Florida Keys area occur along with a small but distinctive fauna of other pleustonic and planktonic mollusks, including the air-filled, surface-dwelling sea slug *Glaucus atlanticus* (the "Blue Glaucus"), the Paper Argonaut cephalopods *Argonauta argo* and *Argonauta hians*, and the Glassy Nautilus, *Carinaria lamarcki* (actually not a nautilus but a swimming heteropod gastropod). Like the janthinids, the Blue Glaucus feeds on the siphonophores, biting off pieces of the stinging tentacles and swallowing them with impunity. The largest and most frequently encountered species of Violet Snail, *Janthina janthina* (Figure 9.8A, B), has a distinctly two-toned shell, with a dark purple base and pale blue-white spire (the namesake of the *Janthina janthina* Assemblage). When the living animal is suspended upside-down from its bubble raft, the shell is perfectly camouflaged: the pale blue spire blends in with the silvery sea surface and protects the snail from predatory fish by making the shell difficult to see from below; the deep purple shell base blends in with the deep blue sea surface as seen from above, making it difficult for predatory birds to locate the floating snail. Three other smaller species of janthinids also occur along with *Janthina janthina*: the pale pinkish-purple *Janthina (Violetta) pallida* (Figure 9.8C, D; which feeds primarily on *Velella*); the dark purple *Janthina (Jodina) exigua* (Figure 9.8E, F; which feeds primarily on *Porpita*); and the pale blue-purple *Janthina (Violetta) globosa* (Figure 9.8G; which feeds primarily on *Velella*). Living along with the purple janthinids is the brown-colored janthinid *Recluzia rollandiana* (Figure 9.8H), which feeds on the brown floating sea anemone *Minyas*. This small species is never as abundant as its purple and blue relatives and is only rarely seen or collected. Along the Palm Beach coast, the yearly appearance of the Great Blue Fleet on beaches during the windy winter time is greatly anticipated by local shell collectors and marine biologists.

Biodiversity of the Janthina janthina Assemblage

Some of the more important pleustonic shelled gastropods of the Florida Keys and adjacent coasts include the following:

CARNIVORES (siphonophore hydrozoan and pelagic zoantherian feeders)
    Gastropoda (pleuston)
    Janthinidae
        Janthina janthina (Linnaeus, 1758) (Figure 9.8A, B)
        Janthina (Jodina) exigua Lamarck, 1816 (Figure 9.8E, F)
        Janthina (Violetta) globosa Swainson, 1822 (Figure 9.8G)
        Janthina (Violetta) pallida (Thompson, 1840) (Figure 9.8C, D)
        Recluzia rollandiana Petit, 1853 (feeds on the floating zoantherian sea anemone
            Minyas) (Figure 9.8H)

Figure 9.8 Gastropods of the *Janthina janthina* Assemblage. A, B = *Janthina janthina* (Linnaeus, 1758), width 29 mm. C, D = *Janthina (Violetta) pallida* (Thompson, 1840), width 18 mm. E, F = *Janthina (Jodina) exigua* Lamarck, 1816, width 14.4 mm. G = *Janthina (Violetta) globosa* Swainson, 1822, width 14 mm. H = *Recluzia rollandiana* Petit, 1853, length 18 mm.

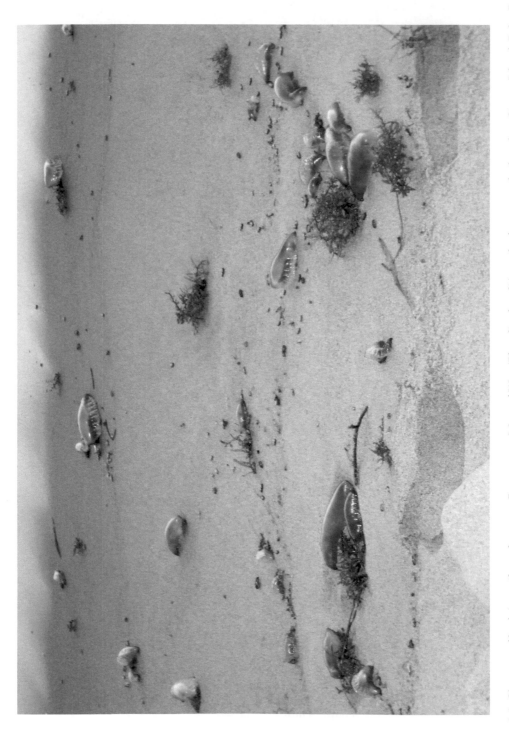

Figure 9.9 Close-up of both juvenile and mature Portuguese Man-of-War (*Physalia physalis*) washed up onto the beach at Singer Island, Palm Beach County. These colonial hydrozoans serve as the principal food resource for the pleustonic gastropod *Janthina janthina* and support the *Janthina janthina* Assemblage of the Open Oceanic Pleustonic Macrohabitat. (Photograph by Jennie Petuch.)

chapter ten

Tree snails of the Florida Keys

Introduction

Although not included within the aquatic regimes and macrohabitats, and technically outside the scope of this book, the tropical hardwood hammocks of the Florida Keys archipelago offer a unique biotope that houses an interesting and highly endemic fauna of tree snails. No book on the malacofaunas of the Florida Keys would be complete without, at least, a short overview of these spectacular endemic gastropods. The same biotic principles of geographical isolation that have produced the endemic marine malacofauna have also worked in the insular terrestrial environment, allowing for the evolution of extremely rich tree snail faunas on many of the widely separated island groups. On most of the larger keys, from Elliott Key south to Key West, dense tropical hardwood forests had become established after the last major Pleistocene sea-level rise (during the Sangamonian Stage; see Petuch and Roberts, 2007; Petuch and Sargent, 2011c). These floras were derived primarily from Caribbean and West Indian taxa, with the incipient forests having been brought to the newly formed islands by seeds in bird droppings or from seeds and propagules that washed ashore during hurricanes. The primary source for these tropical hardwood tree forests was the continental-type island of Cuba, only 120 km to the south. The newly formed late Pleistocene Keys archipelago thus became the gateway into North America for tropical migrants, both plant and animal, from the West Indies and Caribbean Basin.

Typically, the Florida Keys hardwood hammocks are made up of between 15 and 20 different types of trees, with almost all of them classic Caribbean Basin species. Of the more commonly encountered types, only two are endemic to the Florida Keys: the Soldierwood Tree (*Colubrina elliptica*), which is restricted to the hammocks of the Upper Keys, and the Rough Strongback Tree (*Bourreria radula*), which is restricted to the hammocks of the Lower Keys. Some of the more abundant types of hardwoods found in the Florida Keys include the following:

Canopy Species
   Gumbo Limbo (*Bursera simaruba*)
   Pigeon Plum (*Coccoloba diversifola*)
   Paradise Tree (*Simarouba glauca*)
   Wild Mastic (*Sideroxylon foetidissimum*)
   Willow Bustic (*Sideroxylon salicifolium*)
   Jamaican Dogwood (*Piscidia piscipula*)
Understory Species
   White Stopper (*Eugenia axillaris*)
   Spanish Stopper (*Eugenia foetida*)
   Red Stopper (*Eugenia rhombea*)
   Poisonwood (*Metopium toxiferum*)

Florida Boxwood (*Schaefferia frutescens*)
Lignum Vitae (*Guajacum sanctum*)
West Indian Mahogany (*Swietenia mahagoni*)
Bahama Strongbark (*Bourreria succulenta*)
Wild Tamarind (*Lysiloma latilisiquum*)
Ironwood (*Krugiodendron ferreum*)
Bahama Wild Coffee (*Psychotria ligustrifolia*)
Key Thatch Palmetto (*Leucothrinax morrisii*)

Because the Florida Keys are made up mostly of coral limestone (Key Largo Formation fossil reefs of the Upper and Middle Keys) or oölitic limestone (Miami Formation oölite banks of the Lower Keys), the islands are covered by only a thin veneer of soil. Because of this lack of deep soils, the tropical hardwoods of the Florida Keys are stunted and dwarfed, essentially having undergone a natural "bonsai" process because of the restriction of root growth. The dwarf hardwood forests of the Florida Keys (Figure 10.1) are now known to house one of the most beautiful and distinctive tree snail faunas found anywhere in the tropical Americas. This unique terrestrial malacofauna is composed of multiple species in the genera *Liguus* and *Orthalicus* (both Orthalicidae) and *Drymaeus* (Bulimulidae), with the two orthalicid genera comprising taxa found only in the Florida Keys and the Everglades region. Here, these arboreal gastropods feed on thick growths of lichens and algae that cover the bark of trees such as Spanish Stopper, Poisonwood, Pigeon Plum, Wild Tamarind, Ironwood, and Jamaican Dogwood. All three of these arboreal genera have closely related congeners in Cuba, and several represent subspecific offshoots of well-known Cuban forest species. As postulated by Pilsbry (1946: 46–49), the Florida Keys tree snails arrived from Cuba during the late Pleistocene by rafting on large tree and vegetation mats that would have washed down rain-swollen rivers during hurricanes. Strong northward-blowing storm winds would have rapidly pushed the tree rafts across the narrow Florida Straits and deposited them, intact, on the shorelines of the newly formed Florida Keys.

Of the tree snail groups found in the Keys, the orthalicid genus *Liguus* is the most morphologically diverse, being represented by a large swarm of color forms and subspecies. All of these are closely related to the Cuban Banded Tree Snail, *Liguus fasciatus* (Müller, 1774), and are considered to be endemic Floridian subspecies. As outlined in the classification scheme of Pilsbry (1946: 37–102), eight subspecies of *Liguus fasciatus fasciatus* are recognized in the Florida Keys and the Everglades region:

Liguus fasciatus graphicus Pilsbry, 1912 (with 9 named forms)
Liguus fasciatus solidus (Say, 1825) (with 4 named forms)
Liguus fasciatus matecumbensis Pilsbry, 1912 (with 1 named form)
Liguus fasciatus castaneozonatus Pilsbry, 1912 (with 11 named forms)
Liguus fasciatus elliottensis Pilsbry, 1912 (with 4 named forms)
Liguus fasciatus testudineus Pilsbry, 1912 (with 16 named forms)
Liguus fasciatus lossmanicus Pilsbry, 1912 (with 3 named forms)
Liguus fasciatus septentrionalis Pilsbry, 1912

Of these eight subspecies, only *Liguus fasciatus septentrionalis* is not present in the Florida Keys. This northernmost *Liguus* subspecies is, instead, restricted to the mainland Atlantic Coastal Ridge hammocks of Palm Beach and Broward Counties.

The named forms of and subspecies of *Liguus* represent isolated populations that exhibit distinctive and consistent differences in shell morphology. These differences take into account shell shape and size, types of color patterns on the shells, and colors of the protoconchs. As discovered by the first malacological explorers and tree snail collectors in the Florida Keys, each key and island group had its own endemic swarm of color forms, the result of genetic isolation on isolated land masses. The intervening saltwater channels between the islands acted as barriers to gene flow, and many of these distinctive subspecies and forms (several illustrated further in this chapter) undoubtedly represent examples of incipient speciation. As the Keys became developed (and in many cases, overdeveloped) during the years since the 1960s, many of the host hammocks were destroyed by housing construction, and several of the forms and subspecies became locally extinct (extirpated). Fortunately, several local tree snail collectors, particularly the late Archie Jones of Miami, have managed to rescue several populations of endangered Keys *Liguus* forms and move them to isolated hammocks within Everglades National Park. Here, the gene pools that evolved in the Florida Keys can flourish in new habitats on the Florida mainland, preserving for posterity the record of evolution that was once active on the islands to the south. Because of these rescue attempts, however, many of the forms and subspecies now live together in strange and different combinations, and new hybrids and variants are now forming in many of the transplant hammocks. According to the ICZN (International Commission on Zoological Nomenclature), all of these named color forms have no real taxonomic status, and they are only to be considered informal nicknames for distinctive populations.

As pointed out by Pilsbry (1946: 46–49), several introductions of *Liguus* stock from Cuba took place during the late Pleistocene. Based on both shell morphology and biogeographical distribution, the oldest immigrant subspecies appears to be *Liguus fasciatus septentrionalis* of the mainland of southeastern Florida. A more recent wave of migration, probably from north-central Cuba (based on morphological relationships to forms in Matanzas and Santa Clara Provinces) is the complex of forms seen in *Liguus fasciatus castaneozonatus* and *Liguus fasciatus elliottensis*. Based on shell morphology and biogeographical range in the Florida Keys, *Liguus fasciatus graphicus* and *Liguus fasciatus solidus* (especially the form *pictus*) appear to be the most recent immigrants from Cuba, showing close morphological relationships with *Liguus fasciatus* subspecies in Mariel, Cabanas, Puerto Esperanza, and other areas in west-central Cuba. The endemic subspecies *testudineus, matecumbensis,* and *lossmanicus* most probably represent local evolutionary events that resulted from the alternating isolation and conjoining of gene pools during sea-level fluctuations in the latest Pleistocene.

During Pleistocene sea-level lows, many of the keys were united into single large land masses that were covered by hardwood hammocks, allowing for free gene flow within the resident *Liguus* populations. During Pleistocene sea-level highs, these large islands were dissected into smaller land masses, and the original single ancestral tree snail population was broken into several smaller isolated gene pools. This alternating pattern of the formation of single large land masses and subsequent groups of smaller islands led to the evolution of the striking color forms seen on many of the present-day Keys island chains. A classic example of this type of incipient subspecies formation in the Florida Keys is seen in the *Liguus fasciatus graphicus* group of the Middle Keys, where the morphologically closely related, yet distinct, forms *innominatus, lignumvitae, osmenti, vonpaulseni, delicatus, dryas,* and *simpsoni* all occur on different islands in an area extending from Big Pine Key to Lower Matecumbe Key.

Of the 55 named *Liguus* subspecies and color forms, only 27 presently occur on the Florida Keys island chain. Three of the eight named subspecies (*graphicus, solidus,* and *matecumbensis* and all their forms) are the only taxa that originally were restricted to the Florida Keys. These now occur on some of the mainland hammocks because of the

transplanting of populations for rescue purposes. The other four subspecies found in the Keys (*castaneozonatus, elliottensis, testudineus,* and *lossmanicus*) all naturally occur on the Florida mainland in an area centered on the Everglades National Park and the Atlantic Coastal Ridge of Dade, Broward, and Palm Beach Counties. Of these, *testudineus* has the largest number of mainland forms (16), with only 2 (*marmoratus* and *ornatus*) found in the Upper and Middle Keys. Similarly, the subspecies *castaneozonatus* has 11 named forms, but only 3 (typically *castaneozonatus, roseatus,* and *lineolatus*) are found in the Upper Keys (*L. fasciatus castaneozonatus* form *roseatus* shown on Figure 10.2). Only the Florida Keys *Liguus* populations are discussed and illustrated in this book.

Liguus *tree snails of the Upper Florida Keys*

Based on the inferred patterns of migration from Cuba, shell morphology, and present-day biogeographical distribution, the seven Florida Keys *Liguus* subspecies break into two distinct groups. The first of these, the tree snails of the Upper Keys, comprise four subspecies and six named forms. For the most part, these *Liguus* are confined to an area extending from Elliott and Totten Key south to Key Largo. In the hammocks of northern Key Largo, especially in the Card Sound Road area, *Liguus fasciatus castaneozonatus* (Figure 10.3A) and its forms *roseatus* (Figure 10.3B) and *lineolatus* (Figure 10.3C) are the most frequently encountered types. These are often abundant on roadside exposures of the dwarf hardwood hammocks and can readily be seen from a moving vehicle. Unlike the Middle and Lower Keys subspecies, the northern Key Largo *Liguus* specimens are relatively small for the species, usually averaging about half to two-thirds the length of their congeners farther south. Considering the abundance of individual tree snails in northern Key Largo, this dwarfing may be caused by the heavy competition for lichen food resources on the bark of the resident tropical hardwood trees. Besides smaller sizes, the northern *Liguus* forms are also much duller than their highly polished relatives in the Middle and Lower Keys, having a distinct matte finish to their shells.

The Upper Keys *Liguus* populations also include several endemic forms of the subspecies *lossmanicus, testudineus,* and *elliottensis*. The first-mentioned subspecies, in its typical form (Figure 10.3G), was originally thought to be confined to Lossman's Key near Cape Sable but has now been found in areas near Miami and in the southern Everglades hammocks. The *lossmanicus* color forms *luteus* and *aurantius* (Figure 10.3H) have been collected on Key Largo and Vaca Key, demonstrating that the subspecies is also present in the Upper and Middle Keys. The similar-appearing *Liguus fasciatus elliottensis*, with its characteristic green stripes (Figure 10.3D), is found throughout the northernmost Keys, particularly Elliott Key, Totten Key, and northernmost Key Largo. Intensely colored black-banded forms of the subspecies *castaneozonatus* (shown here in Figure 10.3A) are also known from the northernmost hammocks on Key Largo, where they occur together with the bright

Figure 10.1 View of a dwarf hardwood hammock on No Name Key, Lower Florida Keys. Throughout the Florida Keys, forests like this one typically contain groves of tropical trees such as Spanish Stopper (*Eugenia foetida*), Red Stopper (*Eugenia rhombea*), White Stopper (*Eugenia axillaris*), Ironwood (*Krugiodendron ferreum*), Florida Boxwood (*Schaefferia frutescens*), and Pigeon Plum (*Coccoloba diversifolia*). These dense tropical forests, isolated on widely separated islands, provide the habitats for a large fauna of endemic tree snails. This hammock also shelters a small herd of Key Deer (*Odocoileus virginianus clavium*), an endemic Florida Keys subspecies of the White-Tailed Deer that lives only on 26 islands near Big Pine Key and No Name Key. (Photograph by Robert Myers.)

Figure 10.2 Close-up of a living *Liguus fasciatus castaneozonatus* Pilsbry, 1912 (form *roseatus* Pilsbry, 1912), crawling on a tree branch in a dwarf hardwood hammock on the northern end of Key Largo, Upper Florida Keys. (Photograph by Robert Myers.)

Figure 10.3 *Liguus* Tree Snails of the Upper Florida Keys. A = *Liguus fasciatus castaneozonatus* Pilsbry, 1912, typical form, length 43 mm, northern Key Largo. B = *Liguus fasciatus castaneozonatus* Pilsbry, 1912 (form *roseatus* Pilsbry, 1912), length 40 mm, northern Key Largo. C = *Liguus fasciatus castane-ozonatus* Pilsbry, 1912 (form *lineolatus* Simpson, 1920), length 45 mm, northern Key Largo. D = *Liguus fasciatus elliottensis* Pilsbry, 1912, typical form, length 40 mm, Elliott Key. E = *Liguus fasciatus testu-dineus* Pilsbry, 1912 (form *marmoratus* Pilsbry, 1912), length 51 mm, southern Key Largo. F = *Liguus fasciatus testudineus* Pilsbry, 1912 (form *ornatus* Simpson, 1920), length 39 mm, northern Key Largo. G = *Liguus fasciatus lossmanicus* Pilsbry, 1912, typical form, length 46 mm, Lossman's Key. H = *Liguus fasciatus lossmanicus* Pilsbry, 1912 (form *aurantius* Clench, 1929), length 38 mm, northern Key Largo.

red-orange color form of *Liguus fasciatus testudineus*. This particularly beautiful shell, the form *ornatus* (Figure 10.3F), is found only on Totten Key and northern Key Largo and is one of the most striking endemic color variants found in the Upper Keys. The typical dark-colored variety of the subspecies *testudineus*, the form *marmoratus* (Figure 10.3E), is found in large hammocks from southern Key Largo to Vaca Key but is also a commonly encoun-tered species in many of the isolated hammocks ("tree islands") of the southern areas of Everglades National Park.

The nine *Liguus fasciatus* subspecies and forms known from the dwarf hardwood ham-mocks of the Upper Florida Keys (Elliott Key south to Key Largo and Lossman's Key near Cape Sable) are listed below:

Family Orthalicidae
> *Liguus fasciatus castaneozonatus* Pilsbry, 1912 (typical dark banded form; Figure 10.3A)
>> form *roseatus* Pilsbry, 1912 (Figure 10.3B)
>> form *lineolatus* Simpson, 1920 (Figure 10.3C)
> *Liguus fasciatus elliottensis* Pilsbry, 1912 (typical form; Figure 10.3D)
> *Liguus fasciatus testudineus* Pilsbry, 1912 (typical form not found in the Keys; only
>> mainland Florida and the Everglades National Park)
>> form *marmoratus* Pilsbry, 1912 (Figure 10.3E)
>> form *ornatus* Simpson, 1920 (Figure 10.3F)
> *Liguus fasciatus lossmanicus* Pilsbry, 1912 (Figure 10.3G)
>> form *luteus* Simpson, 1920
>> form *aurantius* Clench, 1929 (Figure 10.3H)

Liguus *tree snails of the Middle and Lower Florida Keys*

With the exception of *elliottensis* form *vacaensis* and *testudineus* form *marmoratus*, only three *fasciatus* subspecies and their color forms make up the entire *Liguus* fauna of the Middle and Lower Florida Keys. These subspecies—*fasciatus graphicus, fasciatus solidus,* and *fasciatus matecumbensis*—have developed 14 different forms between them, and these are (or were originally) confined to single islands in the central and lower archipelagos. Here, in the dense and often cactus-filled hammocks of the southern island chains, *Liguus* snails grow to much larger sizes than they do in the hammocks of the northern Keys (Figure 10.4). Because of the recent runaway housing development and cutting down of the large hardwood hammocks on most of the islands, however, many of the most beau-tiful forms are now extinct in their original habitats. One of the areas to experience the devastating impact of this unchecked housing development is the Simpson Hammock on Lower Matecumbe Key, originally the habitat of one of the richest and most spectacu-lar *Liguus* faunas in the Keys. Here, six different endemic forms once occurred together, including the *dohertyi, delicatus,* and *simpsoni* forms of the subspecies *graphicus*; the *pseu-dopictus* (Figure 10.5F) and *splendidus* forms of the subspecies *solidus*; and the *subcrenatus* form of the subspecies *matecumbensis*. The Lower Matecumbe hammocks also experienced an even more devastating force in 1935, when a huge storm surge from an exceptionally strong hurricane destroyed much of the original tropical hardwood-associated fauna and flora. This natural catastrophe led to the extinction of the Simpson Hammock endemic *Liguus fasciatus graphicus* form *dohertyi*. The six remaining endemic Lower Matecumbe forms were able to survive until the 1980s, when the destruction of the hammocks by housing construction began in earnest.

Figure 10.4 Close-up of a living *Liguus fasciatus graphicus* Pilsbry, 1912 (form *innominatus* Pilsbry, 1930), crawling on a tree trunk in a dwarf hardwood hammock on No Name Key, Lower Florida Keys. (Photograph by Robert Myers.)

Of the three main subspecies found in the Middle and Lower Florida Keys, *Liguus fasciatus graphicus* (Figure 10.5A) is the most prominent, having evolved nine different forms. Members of this complex of highly colorful and morphologically similar shells, all obviously having evolved from a common ancestor in the recent past, are restricted to widely separated islands. Classic *graphicus*, itself, is found only on Little Pine and Big Pine Keys (locally extirpated), while the other forms are endemic to individual islands in an arc extending from Lower Matecumbe Key to Middle Torch Key. Some of the more prominent and strikingly beautiful *graphicus* insular variants include the form *innominatus* (endemic to No Name Key; Figure 10.5B), the form *lignumvitae* (endemic to Lignumvitae Key and the southernmost tip of Lower Matecumbe Key (extirpated?) (Figure 10.5C); the form *osmenti* (endemic to Howe Key; Figure 10.5D); and the form *dryas* (endemic to Little Pine Key).

The subspecies *fasciatus solidus* appears to have been the hardest hit by the ecological and habitat damage caused by housing development. Of the four named forms of *solidus*, two are now thought to be extinct (form *solidulus* and the nominate form *solidus*, both originally from Stock Island and Key West). The classic form *pictus* (Figure 10.5E) is extirpated from its home hammocks on Big Pine Key but survives in a few isolated hammocks in Everglades National Park. Likewise, the two Lower Matecumbe Key endemic forms of the subspecies *solidus*, *pseudopictus* (Figure 10.5F) and *splendidus*, are now highly endangered and are extremely rare in their original habitat at Simpson Hammock. Only *Liguus fasciatus matecumbensis* (Figure 10.5G) appears to be maintaining stable colonies in the hammocks on Upper Matecumbe Key. A distinctive pale color variety of *fasciatus elliottensis*, the form *vacaensis* (Figure 10.5H), is found primarily on Vaca Key, where it is frequently encountered in the Crane Point Hammock at Marathon. Very large variants of the form *vacaensis* have also recently been collected on No Name Key near Big Pine Key, and this collecting site would constitute the farthest south record for the members of the *elliottensis* complex.

The four subspecies and 14 named forms known from the dwarf hardwood hammocks of the Middle and Lower Florida Keys (Upper Matecumbe Key south to Key West) are as follows:

Family Orthalicidae
Liguus fasciatus graphicus Pilsbry, 1912 (typical form; Figure 10.5A)
form *innominatus* Pilsbry, 1930 (endemic to No Name Key; Figure 10.5B)
form *lignumvitae* Pilsbry, 1912 (endemic to Lignumvitae Key; Figure 10.5C)
form *osmenti* Clench, 1942 (endemic to Howe Key; Figure 10.5D)
form *dryas* Pilsbry, 1932 (endemic to Little Pine Key)

Figure 10.5 Liguus Tree Snails of the Middle and Lower Florida Keys. A = *Liguus fasciatus graphicus* Pilsbry, 1912, typical form, length 52 mm, northern Big Pine Key. B = *Liguus fasciatus graphicus* Pilsbry, 1912 (form *innominatus* Pilsbry, 1930), length 63 mm, No Name Key. C = *Liguus fasciatus graphicus* Pilsbry, 1912 (form *lignumvitae* Pilsbry, 1912), length 61 mm, Lignum Vitae Key. D = *Liguus fasciatus graphicus* Pilsbry, 1912 (form *osmenti* Clench, 1942), length 53 mm, Howe Key. This blue-gray variety lacks the bright yellow tones seen in the other *graphicus* forms. E = *Liguus fasciatus solidus* (Say, 1825) form *pictus* (Reeve, 1842), length 56 mm, northern Big Pine Key. F = *Liguus fasciatus solidus* (Say, 1825) (form *pseudopictus* Simpson, 1920), length 60 mm, Lower Matecumbe Key. G = *Liguus fasciatus matecumbensis* Pilsbry, 1912, typical form, length 52 mm, southern Upper Matecumbe Key. H = *Liguus fasciatus elliottensis* Pilsbry, 1912 (form *vacaensis* Simpson, 1920), length 44 mm, Crane Point Hammock, Vaca Key.

form *vonpaulseni* Young, 1960 (endemic to Middle Torch Key)
form *crassus* Simpson, 1920 (endemic to Howe Key)
form *delicatus* Simpson, 1920 (endemic to Lignumvitae and Lower Matecumbe Keys)
form *simpsoni* Pilsbry, 1921 (endemic to Lower Matecumbe Key)
Liguus fasciatus solidus (Say, 1825) (originally endemic to Key West; possibly extinct)
form *solidulus* Pilsbry, 1912 (originally endemic to Stock Island; possibly extinct)
form *pictus* (Reeve, 1842) (endemic to northern Big Pine Key; Figure 10.5E)
form *pseudopictus* Simpson, 1920 (endemic to Lower Matecumbe Key; Figure 10.5F)
form *splendidus* Frampton, 1932
Liguus fasciatus matecumbensis Pilsbry, 1912 (endemic to Upper Matecumbe Key; Figure 10.5G)
form *subcrenatus* Pilsbry, 1912 (endemic to Lower Matecumbe Key)
Liguus fasciatus elliottensis form *vacaensis* Simpson, 1920 (found from southern Key Largo south to Vaca Key and No Name Key; Figure 10.5H).

The *Liguus* forms restricted to the hardwood hammocks of mainland Everglades National Park and adjacent highland areas along the Atlantic Coastal Ridge are listed next. Besides occasional errant specimens that have been transplanted by tree snail collectors, none of these forms occurs naturally in the Florida Keys archipelago. These are listed here simply to demonstrate the overall biodiversity of the *Liguus* snails in southern Florida.

Liguus fasciatus septentrionalis Pilsbry, 1912 (Palm Beach to southern Broward Counties)
Liguus fasciatus castaneozonatus Pilsbry, 1912 (eight endemic mainland forms)
form *walkeri* Clench, 1933 (Pinecrest area, Everglades)
form *miamiensis* Simpson, 1920 (Brickell Hammock, Miami, and southern Dade County)
form *elegans* Simpson, 1920 (Atoll Hammock, Everglades)
form *deckerti* Clench, 1935 (southern Everglades)
form *humesi* Jones, 1979 (southern Everglades)
form *framptoni* Jones, 1979 (southern Everglades)
form *alternatus* Simpson, 1920 (Timm's Hammock and Redlands area, Dade County)
form *livingstoni* Simpson, 1920 (Brickell Hammock, Miami)

Liguus fasciatus elliottensis (three endemic mainland forms)
form *cingulatus* Simpson, 1920 (Brickell Hammock, Miami)
form *eburneus* Simpson, 1920 (southern Everglades)
form *capensis* Simpson, 1920 (Cape Sable, Everglades National Park)

Liguus fasciatus testudineus Pilsbry, 1912 (12 living and 2 extinct endemic mainland forms)
form *castaneus* Simpson, 1920 (southern Everglades)
form *versicolor* Simpson, 1920 (southern Everglades)
form *nancyae* Close, 1994 (Long Pine Key, Everglades National Park)
form *clenchi* Frampton, 1932 (northwestern Everglades)
form *evergladesensis* Jones, 1979 (southern Everglades)
form *fuscoflamellus* Frampton, 1932 (Timm's Hammock and Redlands, Dade County)
form *barbouri* Clench, 1929 (southern Everglades)
form *floridanus* Clench, 1929 (Pinecrest area, Everglades)
form *gloriasylvaticus* Doe, 1937 (Pinecrest area, Everglades)
form *nebulosus* Doe, 1937 (western Everglades, Collier County)
form *solisocassis* deBoe, 1933 (western Everglades, Collier County)

form *lucidovarius* Doe, 1937 (Pinecrest area, Everglades)

extinct form *farnumi* Clench, 1929 (Hammock #7, Pinecrest area, Everglades; destroyed by forest fires and hunter's off-road traffic)

extinct form *violafumosus* Doe, 1937 (Hammock #28 and #30, Pinecrest area, Everglades; destroyed by several forest fires)

Liguus fasciatus lossmanicus (one mainland endemic form)

form *mosieri* Simpson, 1920 (Brickell Hammock, Miami)

Several man-made stable hybrids were created by Archie Jones in the 1970s as deliberate crosses between two or more color forms (discussed by Power, 2010). These distinctive artificial color forms, *margaretae* Jones, 1979, *kennethi* Jones, 1979, *beardi* Jones, 1979, and *wintei* Jones, 1979, were subsequently released into several hammocks in southern Dade County. An excellent overview of all the *Liguus fasciatus* subspecies and forms, with color photographs of every taxon, is given on the *Liguus* of South Florida and the Florida Keys website by Emilio Jorge Power (last update 2010).

Endemic orthalicid and bulimulid tree snails of the Florida Keys

Besides the large and characteristic fauna of *Liguus* subspecies and forms, the dwarf hardwood hammocks of the Florida Keys also house a small fauna of other types of tree snails (Figure 10.6). These include two species and one subspecies of the related genus *Orthalicus*, another orthalicid group that originated in the tropical forests of Cuba and rafted to the Florida Keys during Pleistocene hurricanes. The most widespread of the three named taxa is *Orthalicus floridensis* (Pilsbry, 1899) (Figure 10.7E, F), which ranges throughout the Florida Keys and onto the southern edge of the Florida mainland, from southern Dade County to the southern areas of Everglades National Park in Collier County. The other related species, *Orthalicus reses reses* (Figure 10.7A, B), was originally endemic to Stock Island and Key West but was transplanted to other selected hammocks around the Florida Keys during the 1980s. Threatened with extinction because of habitat loss, workers from the University of Florida moved individuals from the public golf course on Stock Island, the only place where the species was known to have survived, to adjacent keys. From these transplanted colonies, tree snail collectors have now spread *reses* colonies all along the island chain, from southern Key Largo, Big Pine Key, No Name Key, to Monkey Jungle (near Timm's Hammock) in Miami, Dade County. A smaller, heavier, and more colorful subspecies of *Orthalicus reses* was described from the hardwood hammocks of the Middle Florida Keys, particularly Vaca Key and Lower Matecumbe Key. This distinctive subspecies, *Orthalicus reses nesodryas* (Figure 10.7C, D), is now less commonly-encountered than the once almost-extinct *reses reses*.

The third group of tree snails found in the Florida Keys is the genus *Drymaeus*, comprising two species and two forms. Resembling tiny *Liguus* snails (Figure 10.8), these small arboreal gastropods actually belong to the family Bulimulidae, and their morphological similarities to the orthalicid genera are the result of convergent evolution. In the Florida Keys, the most abundant bulimulid species is the brightly patterned *Drymaeus multilineatus* (Figure 10.7G), a widespread species that is also found throughout the Caribbean Basin and Yucatan, Mexico. This fragile species also ranges northward on the Florida mainland to Broward and Palm Beach Counties. A wide-banded subspecies, *Drymaeus multilineatus latizonatus*, was also named from Lower Matecumbe Key, Long Key, and Lignumvitae

Figure 10.6 Close-up of a living *Orthalicus reses reses* (Say, 1830) crawling on a tree branch in a dwarf hardwood hammock on No Name Key, Lower Florida Keys. (Photograph by Robert Myers.)

Figure 10.7 *Orthalicus* and *Drymaeus* Tree Snails of the Florida Keys. A, B = *Orthalicus reses reses* (Say, 1830), length 58 mm, No Name Key (from a transplanted colony; originally endemic to Stock Island and Key West). C, D = *Orthalicus reses nesodryas* Pilsbry, 1946, length 52 mm, Crane Point, Vaca Key. E, F = *Orthalicus floridensis* (Pilsbry, 1899), length 50 mm, Crane Point, Vaca Key. G = *Drymaeus multi-lineatus* (Say, 1825), length 22 mm, No Name Key. H = *Drymaeus hemphilli* (B.H. Wright, 1889), typical form, length 17 mm, northern Key Largo. I = *Drymaeus hemphilli* (B.H. Wright, 1889) (form *clarissimus* Pilsbry, 1946), length 17 mm, Card Sound Road, northern Key Largo. J = *Drymaeus hemphilli* (B.H. Wright, 1889), single-banded color form, length 18 mm.

Figure 10.8 Close-up of a living *Drymaeus multilineatus* (Say, 1825) crawling on a tree branch in a dwarf hardwood hammock on No Name Key, Lower Florida Keys. (Photograph by Robert Myers.)

Key. Further research, particularly DNA studies, may show that the Florida populations of *Drymaeus multilineatus* are distinct from the Caribbean ones and that the taxon may actually represent a complex of closely related taxa. A second bulimulid species, *Drymaeus hemphilli* (Figure 10.7H, J), also occurs in the Florida Keys, being found in small colonies along the northern side of Key Largo and on the eastern side of Upper Matecumbe Key. This colorful banded species is endemic to southern Florida, from Sebring, Highlands County, south to the Middle Florida Keys, and is closely related to *Drymaeus dominicus* from the Greater Antilles Arc. *Drymaeus hemphilli* is another example of an endemic Floridian species that evolved from a Pleistocene Cuban ancestor. The unbanded color form *clarissimus* Pilsbry, 1946 (Figure 10.7I) has also been collected in hammocks along Card Sound Road on northern Key Largo.

Listings of the non-*Liguus* orthalicid and bulimulid tree snails known from the dwarf hardwood hammocks of the Florida Keys (Key Largo to Key West) follow:

Family Orthalicidae
Orthalicus floridensis (Pilsbry, 1899) (Key Largo to Key West) (Figure 10.7E, F)
Orthalicus reses reses (Say, 1825) (originally endemic to Stock Island) (Figure 10.7A, B)
Orthalicus reses nesodryas Pilsbry, 1946 (Middle Keys) (Figure 10.7C, D)
Family Bulimulidae
Drymaeus hemphilli (B.H. Wright, 1889) (Key Largo to the Matecumbe Keys) (Figure 10.7H, J)
form *clarissimus* Pilsbry, 1946 (northern Key Largo) (Figure 10.7I)
Drymaeus multilineatus (Say, 1825) (Key Largo to Key West) (Figure 10.7G)
form *latizonatus* Pilsbry, 1936 (Lower Matecumbe Key to Long Key)

Close-up of a young Key Deer buck (*Odocoileus virginianus clavium*) on the edge of a dwarf hardwood hammock on No Name Key, Lower Florida Keys. A small herd of this Keys endemic subspecies of the White-Tailed Deer occurs in the same hammock that shelters the rare endemic No Name Tree Snail, *Liguus fasciatus graphicus* form *innominatus*.

Bibliography

Abbott, R.T. 1974. *American Seashells* (second edition). Van Nostrand Reinhold, New York. 663 pp.

Abbott, R.T. 1976. *Cittarium pica* (Trochidae) in Florida. *The Nautilus* 90(1): 2–4.

Bartsch, P.D. 1943. A review of some West Atlantic turritid mollusks. *Memorias de la Sociedad Cubana de Historia Natural* 17(2): 81–122, plates 7–15.

Bieler, R. and R.E. Petit. 2011. Catalogue of the Recent and fossil "worm-snail" taxa of the families Vermetidae, Siliquariidae, and Turritellidae (Mollusca: Caenogastropoda). *Zootaxa* 2948: 1–103.

Brewster-Wingard, G.L. and S.E. Ishman. 1999. Historical trends in salinity and substrate in Central Florida Bay: a paleoecological reconstruction using modern analogue data. *Estuaries* 22(2B): 369–383.

Brewster-Wingard, G.L., J.R. Stone, and C.W. Holmes. 2001. Molluscan faunal distribution in Florida Bay, past and present: an integration of down-core and modern data. *Bulletins of American Paleontology, Special Volume* 361:199–231.

Carson, R. 1955. *The Edge of the Sea.* Houghton Mifflin, Cambridge, Massachusetts. 276 pp.

Chiappone, M. and K.M. Sullivan. 1994. Ecological structure and dynamics of nearshore hard-bottom communities in the Florida Keys. *Bulletin of Marine Science* 54(3): 747–756.

Chiappone, M. and K.M. Sullivan. 1996. *Functional Ecology and Ecosystem Trophodynamics. Site Characterization for the Florida Keys National Marine Sanctuary and Environs*, Volume 8. The Nature Conservancy and Florida and Caribbean Marine Conservation Science Center, University of Miami. 112 pp.

Coomans, H.E. 1969. Biological aspects of mangrove molluscs in the West Indies. *Malacologia* 9(1): 79–84.

Dame, R.F. 2011. *Ecology of Marine Bivalves: An Ecosystem Approach* (second edition). CRC Press, Boca Raton, Florida. 283 pp.

Davis, J.H. 1940. The ecology and the geological role of mangroves in Florida. *Papers from the Tortugas Laboratory of the Carnegie Institute of Washington, Publication* 32: 307–412.

Dawes, C.J. 1987. The dynamic sea grasses of the Gulf of Mexico and Florida coasts. *Florida Marine Research Publications* 42: 25–38.

Dunstan, P. 1985. Community structure of reef-building corals in the Florida Keys: Carysfort Reef, Key Largo and Long Key Reef, Dry Tortugas. *Atoll Research Bulletin* 288: 1–27.

Frankovich, T.A. and J.C. Zieman. 1994. Total epiphyte and epiphytic carbonate production on *Thalassia testudinum* across Florida Bay. *Bulletin of Marine Science* 54(3): 679–695.

Ginsburg, R.N. 1953. Intertidal erosion on the Florida Keys. *Bulletin of Marine Science of the Gulf and Caribbean* 3(1): 55–69.

Goldberg, W.M. 1973. The ecology of the coral-octocoral communities off the southeast Florida coast: geomorphology, species composition, and zonation. *Bulletin of Marine Science* 23(3): 465–487.

Hoffmeister, J.E. and H.G. Multer. 1968. Geology and origin of the Florida Keys. *Geological Society of America Bulletin* 79: 1487–1502.

Jaap, W.C. 1984. *The Ecology of the South Florida Coral Reefs: A Community Profile.* U.S. Fish and Wildlife Service, Office of Biological Services, Washington, D.C. FWS/OBS-82/08. 138 pp.

Lee, H.G. and M. Huber. 2012. *Americardia lightbourni* new species and *A. columbella* new species compared to *A. media* (Linnaeus, 1758), *A. speciosa* (A. Adams and Reeve, 1850), and the extinct *A. columba*. *The Nautilus* 126(1): 15–24.

Levy, J.M., M. Chiappone, and K.M. Sullivan. 1996. *Invertebrate Infauna and Epifauna of the Florida Keys and Florida Bay. Site Characterization for the Florida Keys National Marine Sanctuary and Environs*, Volume 5. Nature Conservancy and Florida and Caribbean Marine Conservation Science Center, University of Miami. 166 pp.

Lipe, R. 1991. *Marginellas*. The Shell Store, St. Petersburg, Florida. Privately printed. 40 pp.

Lyons, W.G. and J.F. Quinn, Jr. 1995. Phylum Mollusca (Checklist). Appendix J. Marine and Terrestrial Species and Algae. In *Draft Management Plan/Environmental Impact Statement, Strategy for Stewardship: Florida Keys National Marine Sanctuary*. Volume 3 (Appendices). National Oceanographic and Atmospheric Administration, Washington, D.C., pp. J-10 to J-25.

Madden, C.J. and K.L. Goodwin. 2007. *Ecological Classification of Florida Bay using the Coastal Marine Ecological Classification Standard* (CMECS). Sponsored by the Curtis and Edith Munson Foundation, NatureServe, Arlington, Virginia. 46 pp.

Mikkelsen, P. and R. Bieler. 2008. *Seashells of Southern Florida: Bivalves*. Princeton University Press, Princeton, New Jersey. 503 pp.

Mullins, H.T. and A.C. Neumann. 1979. Geology of the Miami Terrace and its paleo-oceanographic implications. *Marine Geology* 30(3–4): 205–232.

Petuch, E.J. 1974. A new *Terebra* from the coral reef areas off North Carolina. *The Veliger* 17(2): 205–208.

Petuch, E.J. 1987. *New Caribbean Molluscan Faunas*. Coastal Education and Research Foundation, Charlottesville, Virginia. 154 pp.

Petuch, E.J. 1988. *Neogene History of Tropical American Mollusks: Biogeography and Evolutionary Patterns of Tropical Western Atlantic Mollusca*. Coastal Education and Research Foundation, Charlottesville, Virginia. 217 pp.

Petuch, E.J. 1994. *Atlas of Florida Fossil Shells*. Graves Museum of Archaeology and Natural History, Dania, Florida, and Chicago Spectrum Press. 394 pp.

Petuch, E.J. 2004. *Cenozoic Seas: The View from Eastern North America*. CRC Press, Boca Raton, Florida. 308 pp.

Petuch, E.J. 2008. *The Geology of the Florida Keys and Everglades: An Illustrated Guide to Florida's Hidden Beauty*. Thomson Press, Mason, Ohio. 84 pp.

Petuch, E.J. 2013. *Biogeography and Biodiversity of Western Atlantic Mollusks*. CRC Press, Florida. 234 pp.

Petuch, E.J. and M. Drolshagen. 2011. *Compendium of Florida Fossil Shells. Volume 1. Middle Miocene to Late Pleistocene Marine Gastropods; Families Strombidae, Cypraeidae, Ovulidae, Eocypraeidae, Triviidae, Conidae, and Conilthidae*. MdM, Wellington, Florida. 412 pp.

Petuch, E.J. and C.E. Roberts. 2007. *The Geology of the Everglades and Adjacent Areas*. CRC Press, Florida. 212 pp.

Petuch, E.J. and D.M. Sargent. 1986. *Atlas of the Living Olive Shells of the World*. Coastal Education and Research Foundation, Charlottesville, Virginia. 253 pp.

Petuch, E.J., and D.M. Sargent. 2011a. A new member of the *Gradiconus* species complex (Gastropoda: Conidae) from the Florida Keys. *Visaya* (October, 2011): 98–104.

Petuch, E.J. and D.M. Sargent. 2011b. New species of Conidae and Conilithidae (Gastropoda) from the Tropical Americas and Philippines, with notes on some poorly-known Floridian species. *Visaya* (September, 2011): 117–138.

Petuch, E.J. and D.M. Sargent. 2011c. *Rare and Unusual Shells of Southern Florida*. Conch Republic Books, Mount Dora, Florida. 189 pp.

Pilsbry, H.A. 1946. *Land Mollusca of North America*. Volume 2, Part 1. Academy of Natural Sciences of Philadelphia. 1–520 pp.

Porter, J. (Editor). 2001. *The Everglades, Florida Bay, and Coral Reefs of the Florida Keys: An Ecosystem Sourcebook*. CRC Press, Florida. 1024 pp.

Power, E.J. 2010. *Liguus* of South Florida and the Florida Keys. http://www.liguushomepage.com/fligtax.html

Reed, J.K, S.A. Pomponi, D. Weaver, C.K. Paull, and A.E. Wright. 2005. Deep water sinkholes and bioherms of South Florida and the Pourtales Terrace habitat and fauna. *Bulletin of Marine Science* 77(2): 267–296.

Reynolds, L.K., P. Berg, and J.C. Zieman. 2007. Lucinid clam influence on the biogeochemistry of the seagrass *Thalassia testudinum* sediments. *Estuaries and Coasts* 30(3): 482–490.

Roessler, M., A. Cantillo, and J. Garcia-Gomez. 2002. *Biodiversity Study of Southern Biscayne Bay and Card Sound (1968–1973)*. U.S. Department of Commerce, National Oceanic and Atmospheric Administration and Rosenstiel School of Marine and Atmospheric Science, University of Miami. Joint Publication 2002-01, pp. 1–36, Appendices I and II.

Rudnick, D.T., P.B. Ortner, J.A. Browder, and S.M. Davis. 2005. A conceptual ecological model of Florida Bay. *Wetlands* 25(4): 870–883.

Shier, D.E. 1969. Vermetid reefs and coastal development of the Ten Thousand Islands, southwest Florida. *Geological Society of America Bulletin* 80: 485–508.

Shinn, E.A., B.H. Lidz, R.B. Haley, J.H. Hudson, and J.L. Kindiger. 1989. *Reefs of Florida and the Dry Tortugas*. Field Trip Guidebook T-176, 28th International Geological Congress, American Geophysical Union, Washington, D.C. 53 pp.

Trappe, C.A. and G.L. Brewster-Wingard. 2013. *Molluscan Fauna from Core 25B, Whipray Basin, Central Florida Bay, Everglades National Park*. USGS Open-File Report 2001-143. Department of the Interior, Washington, D.C. 24 pp.

Tucker, J.K., 2013. *The Cone Shells of Florida: An Illustrated Key and Review of the Recent Species*. MdM, Wellington, Florida. 155 pp.

Tunnell, J.W., J. Andrews, N.C. Barrera, and F. Moretzsohn. 2010. *Encyclopedia of Texas Seashells*. Texas A&M University Press, College Park. 512 pp.

Turney, W.J. and B.F. Perkins. 1972. *Molluscan Distribution in Florida Bay. Sedimenta III*. Comparative Sedimentology Laboratory, Division of Marine Geology and Geophysics, Rosenstiel School of Marine and Atmospheric Science, University of Miami. 37 pp.

Voss, G.L. 1988. *Coral Reefs of Florida*. Pineapple Press, Sarasota, Florida. 96 pp.

Voss, G.L. and N.A. Voss. 1955. An ecological survey of Soldier Key, Biscayne Bay, Florida. *Bulletin of Marine Science of the Gulf and Caribbean* 5(3): 203–229.

Warmke, G.L. and R.T. Abbott. 1962. *Caribbean Sea Shells*. Livingston, Narberth, Pennsylvania. 348 pp.

White, W.A. 1970. *The Geomorphology of the Florida Peninsula*. Florida Geological Survey, Bulletin 51. Tallahassee, Florida. 185 pp.

Wingard, G.L. and J.W. Hudley. 2011. Application of a weighted-averaging method for determining paleosalinity: a tool for restoration of South Florida's estuaries. *Estuaries and Coasts* (2012) 35: 262–280.

Zischke, J.A. 1973. *An Ecological Guide to the Shallow-Water Marine Communities of Pigeon Key, Florida*. St. Olaf College Press, Northfield, Minnesota. 220 pp.

Systematic appendix: additions to the molluscan biodiversity of the Florida Keys

While compiling the taxonomic data for the ecosystem analyses shown throughout this book, we found that several prominent Florida Keys mollusks were still undescribed. Because this volume was designed as an attempt to take into account the total macromollusk biodiversity of the area, we felt that these overlooked taxa needed to be formally described in a systematic appendix. In total, seven new species and three new subspecies are described here. The holotypes of nine of the new taxa are deposited in the Division of Mollusks, National Museum of Natural History, Smithsonian Institution, Washington, D.C., and bear USNM numbers. One of the new subspecies is deposited in the Los Angeles County Museum of Natural History and bears an LACM number. The new descriptions of the species and subspecies are given next.

Gastropoda

Turritellidae
Vermicularia fargoi owensi Petuch and Myers, new subspecies (USNM1231410)
Modulidae
Modulus calusa foxhalli Petuch and Myers, new subspecies (LACM3279; Los Angeles County Museum of Natural History)
Naticidae
Neverita delessertiana patriceae Petuch and Myers, new subspecies (USNM1231404)
Muricidae
Murexiella caitlinae Petuch and Myers, new species (USNM1231409)
Buccinidae
Hesperisternia sulzyckii Petuch and Myers, new species (USNM1231411)
Nassariidae
Uzita swearingeni Petuch and Myers, new species (USNM1231412)
Olividae
Americoliva matchetti Petuch and Myers, new species (USNM1231405)
Americoliva recourti Petuch and Myers, new species (USNM1231407)

Bivalvia

Arcidae
  Arca rachelcarsonae Petuch and Myers, new species (USNM1231406)
Pectinidae
  Euvola marshallae Petuch and Myers, new species (USNM1231408)

Descriptions of new taxa

Gastropoda
Prosobranchia
Cerithiodea
Turritellidae
Genus *Vermicularia* Lamarck, 1799

Vermicularia fargoi owensi new subspecies
(Figure 7.14F, G)

> **Description:** Shell small for species, tightly coiled, more turritelloid than nominate subspecies; shell color white to pale tan or straw colored, with small, evenly spaced pale tan dots on cords of later whorls; early whorls of turritelloid section white; early whorls proportionally more slender than those of nominate subspecies; turritelloid section of body whorl comprising almost entire shell length, with only last one or two whorls becoming uncoiled; spire whorls (turritelloid section) ornamented with two faint, thin cords and numerous fine spiral threads and with single large, prominent spiral cord present along peripheral edge near suture, producing distinct pagodiform appearance when viewed in profile; last two whorls ornamented with three large, prominent spiral cords, with anterior-most two cords larger than posterior subsutural cord; aperture slightly rectangular; protoconch white, rounded, and dome-like, composed of two whorls.
>
> **Holotype:** Length 23 mm, width 10 mm, USNM1231410.
>
> **Type Locality:** In muddy sand at low tide, near the vermetid worm shell reef along the northern side of Demijohn Key, Ten Thousand Islands, Collier County, Florida.
>
> **Etymology:** Named for Robert Owens, Boca Raton, Florida, who assisted the senior author in surveying the mollusks of the Ten Thousand Islands.
>
> **Discussion:** Of the three known subspecies of *Vermicularia fargoi* Olsson, 1951, this new endemic Ten Thousand Islands turritellid is closest to the nominate subspecies from western Florida but differs in being a much smaller shell with a proportionally more slender spire. Typically, *Vermicularia fargoi fargoi* has a uniform dark brown body whorl, with the turritelloid early whorls being tan or brown (Abbott, 1974: 96). In *Vermicularia fargoi owensi*, the body whorl is white or pale whitish-tan, and the early whorls are uniformly white or pale straw colored. The spire whorls of the new Ten Thousand Islands subspecies are also different in appearance, being distinctly pagodiform, with a single large spiral cord along the suture that projects beyond the slope of the spire whorls. The new subspecies also does not uncoil as much as the nominate subspecies, instead retaining the turritelloid appearance for most of the shell length.

As far as is presently known, *Vermicularia fargoi owensi* is restricted to the sand-bottomed lagoons behind the vermetid reefs of the outermost Ten Thousand Islands. A third, apparently unnamed, subspecies of *Vermicularia fargoi* occurs in the coastal lagoons of Texas. This unusual western Gulf of Mexico subspecies differs from the eastern Gulf of Mexico *V. fargoi fargoi* and *V. fargoi owensi* by having proportionally much more slender and more protracted turritelloid spire whorls (for an illustration of this unnamed Texas subspecies, see Tunnell et al., 2010: 135; for an overview of the systematics of the worm shell families Vermetidae, Turritellidae, and Siliquariidae, see Bieler and Petit, 2011).

Modulidae
Genus *Modulus* Gray, 1840

Modulus calusa foxhalli new subspecies
(Figure 6.13 H)

Description: Shell of average size for genus and nominate subspecies, trochoidal, with high elevated spire and sloping spire whorls; spire often protracted and scalariform, with detached whorls, producing pagoda-like appearance; periphery of mid-body body and spire whorls with single large, prominent, smooth spiral cord; peripheral cord often ornamented with closely-packed, low, rounded beads (as on holotype); sloping subsutural area of body whorl ornamented with 4 to 5 large, smooth spiral cords; base of shell ornamented with 6 large, prominent spiral cords; umbilicus large, open, completely perforate; columella with single large, bladelike tooth at anterior end; shell colored pale tannish-white or cream-tan, with spiral cords of body whorl and spire being marked with small, evenly-spaced dark brown spots; columellar tooth and columellar area pale purple; aperture white, nearly circular, with 6-8 large cords within interior.

Holotype: width 12 mm, height 11 mm, LACM 3279 (Department of Malacology, Los Angeles County Museum of Natural History, Los Angeles, California); Paratype: width 12 mm, height 12 mm, from the same locality as the holotype, in the research collection of the senior author.

Type Locality: Collected in 0.5 m depth near a large Turtle Grass bed off Pine Point, Singer Island, Lake Worth Lagoon, northern Palm Beach County, Florida.

Etymology: Named for Emyr Foxhall of Cardiff, Wales (Caerdydd, Cymru), who assisted the senior author in collecting mollusks in the Florida Keys.

Discussion: The new taxon is named as an isolated, northern subspecies of the Florida Keys, Florida Bay, and Biscayne Bay *Modulus calusa* Petuch, 1988 (Figure 4.5, Chapter 4). *Modulus calusa foxhalli* differs from the nominate subspecies in being a much more highly-sculptured and ornamented shell, with strong spiral cords on the subsutural area, and in having a much higher, more protracted spire, producing highly-sloped spire whorls. The nominate subspecies, *M. calusa calusa*, typically has low, flattened whorls, with a distinctly discoidal shape, and has a smooth, or only slightly corded, subsutural area. Many specimens of *M. calusa foxhalli* from Lake Worth Lagoon and Peanut Island (West Palm Beach) areas have highly protracted, detached spire whorls, creating a distinctly pagoda-shaped shell. The new subspecies ranges from the Lake Worth Lagoon System of Palm Beach County northward to the

Indian River Lagoon System of Indian River and St. Lucie Counties, and is restricted to quiet coastal lagoons and inlets.

Naticoidea
Naticidae
Polinicinae
Genus *Neverita* Risso, 1826

Neverita delessertiana patriceae new subspecies
(Figure 6.12G–I)

Description: Shell of average size for genus, heavy, thickened, distinctly coni- cal in shape, with high, elevated spire; spire whorls sloping, subpyramidal, faintly shouldered below subsutural area; base of shell characteristically flat- tened; shell color two toned, with spire and posterior two-thirds of body whorl pale tan and gray, and with base of shell being white with scattered pale brown growth lines; darker brown band present along spire suture, espe- cially on early whorls; umbilicus widely open, with umbilical callus covering less than one-half the umbilical opening; umbilical callus tannish-brown, adherent to body whorl only along extreme posterior end; umbilical opening distinctly crescent shaped; interior of aperture dark brown with small whit- ish area at anterior end.

Holotype: Height 52 mm, width 51 mm, USNM1231404; two paratypes, heights 51 mm (in the collection of Patrice Marker, Wellington, FL) and 56 mm (in the research collection of the senior author), both from the same locality as the holotype.

Type Locality: In muddy sand at low tide, on exposed flat north of Manalapan Island, Lake Worth Lagoon, Palm Beach County, Florida.

Etymology: Named for Patrice Marker of Wellington, Florida, life partner of the junior author and well-known underwater photographer and marine naturalist.

Discussion: This new taxon is proposed as a southeastern Floridian subspecies of the large False Shark Eye, *Neverita delessertiana* (Recluz, 1843) (Figure 7.16E– G), which ranges along the northern Gulf of Mexico from Texas to Louisiana, Mississippi, Alabama, and western Florida. *Neverita delessertiana patriceae* dif- fers from the low-spired, globose nominate subspecies in being a distinctly much more conical shell with a much higher, elevated spire and much more flattened shell base. The umbilical opening of the new subspecies also dif- fers from that of the nominate subspecies in being proportionally wider and larger and in having a smaller umbilical callus. The umbilical callus of *N. delessertiana delessertiana* covers approximately one-half the umbilical open- ing and is attached to the body whorl for half its width, producing a dis- tinct trigonal shape. The umbilical callus of *N. delessertiana patriceae*, on the other hand, is attached to the body whorl for only a small part of its length, and the umbilical opening curls around to form a distinct crescent shape. The new subspecies has the smallest umbilical callus and the widest, deep- est, and most open umbilical area of any known western Atlantic *Neverita* species or subspecies. Specimens of *Neverita delessertiana* reported from the Palm Beach coast and southeastern Florida are now known to belong to this new subspecies. For color illustrations of a typical Texas specimen of *Neverita delessertiana delessertiana*, see Tunnel et al., 2007: 177.

Muricoidea
Muricidae
Muricopsinae
Genus *Murexiella* Clench and Farfante, 1945

Murexiella caitlinae new species
(Figure 5.2I, J)

Description: Shell of average size for genus, with inflated body whorl, proportionally long siphonal canal, and high stepped spire; shoulder sharply angled, with flattened subsutural areas, producing scalariform spire whorls; body whorl with five raised varices; areas between varices sculptured with four large, thick, and prominent spiral cords; fifth smaller cord present around juncture of body whorl and siphonal canal; varices sculptured with four large recurved spines and one small spine at base of varix; spire whorls with very thin, small cord present along sutural area; siphonal canal with three large flattened spines on each varix; intervarical ribs and varical spines ornamented with large fluted scales, giving shell distinct squamous appearance; anterior tip of siphonal canal recurved dorsally; aperture oval; parietal shield erect, disconnected from columella; shell color pale cream-white with three pale orange bands, one on flattened subsutural area, one on posterior-most rib, and one on third rib from shoulder; interior of aperture white.

Holotype: Length 30 mm, width 19 mm, USNM1231409.

Type Locality: In a Yellow Mussel bed near Turtle Grass, 1.5-m depth, off the eastern side of Missouri Key, Middle Florida Keys, Monroe County, Florida.

Etymology: Named for Caitlin Hanley, Department of Geosciences, Florida Atlantic University, who collected the holotype off Missouri Key.

Discussion: This large and distinctive muricid has been recognized as occurring in the Florida Keys for many decades now (i.e., Abbott, 1974: 175) but has always been incorrectly referred to as *Murexiella mcgintyi* (M. Smith, 1938). That taxon, although similar to *M. caitlinae*, is actually a 2-million-year-old early Pleistocene (Gelasian Age) fossil named from the Caloosahatchee Formation of southern Florida (see Petuch, 1994: 138, Plate 48, Figures A, B; Petuch and Roberts, 2007: 119–137). The fossil species, although congeneric and ancestral, differs from the living *M. caitlinae* by having a much more globose and inflated body whorl with a distinctly rounded shoulder, by having five major cords on the body whorl and not four as in the living Keys species, and by having more varices per whorl (generally six or seven). The real *M. mcgintyi* became extinct by the Aftonian Stage of the Pleistocene and was replaced by its descendant species, *M. graceae* (McGinty, 1940), during the deposition of the Bermont Formation (mid-Pleistocene; see Petuch and Roberts, 2007: 150–172). *Murexiella graceae*, with its characteristic angled shoulder, is morphologically much more similar to *M. caitlinae* and appears to be the direct ancestor of the living species. The new species appears to be endemic to the Florida Keys, where it occurs on beds of the Yellow Mussel *Brachidontes modiolus*.

Buccinoidea
Buccinidae
Genus *Hesperisternia* Gardner, 1944

Hesperisternia sulzyckii new species
(Figure 9.3E, F)

Description: Shell of average size for genus, elongated, slender, with high stepped spire; siphonal canal proportionally short and stubby, open; body whorl and spire whorls with eight large, sharp-edged axial varices per whorl; shoulder sharply angled, with sloping subsutural area; body whorl ornamented with four large, prominent spiral cords around midsection; three smaller secondary cords between each pair of large cords; intersection of four large spiral cords with varices producing large, elongated knobs and rounded beads; knobs especially prominent and sharp along shoulder cord; sloping subsutural area with six small spiral cords and threads; siphonal canal ornamented with four large spiral cords and numerous small spiral threads; subsutural area depressed between varices, creating series of deep subsutural pits; shell color bright canary yellow with scattered small light brown patches and dots; aperture oval, proportionally large, with 12 large cords on interior and inside of lip; interior of aperture pure white; columella smooth, white in color; protoconch proportionally large, rounded, dome-like, composed of two whorls.

Holotype: Length 26.3 mm, width 12 mm, USNM1231411.

Type Locality: Collected in a lobster trap, from 50-m depth in red algae beds, 10 km north of Garden Key, Dry Tortugas, Florida Keys, Monroe County, Florida.

Etymology: Named for John Sulzycki, senior editor for life sciences at CRC Press, who guided the senior author through the complex publication process of five different books.

Discussion: This new offshore buccinid was illustrated in one of the senior author's previous books on the molluscan faunas of the Florida Keys (Petuch and Sargent, 2011: 101, Figure 4.8B) but was misidentified as *Hesperisternia grandanus* Abbott, 1986. That congeneric taxon is actually a very different-appearing species that is endemic to northwestern Florida and the Florida Panhandle (illustrated in Petuch, 2013: 60, Figure 4.11F). The true *Hesperisternia grandana* differs from the Dry Tortugas endemic *H. sulzyckii* in being a much more inflated shell with a rounded shoulder and body whorl, in lacking strong varices and knobbed shell sculpture on the body whorl, and in being ornamented only with spiral threads and small cords. Although both shells have high, protracted spires, *H. sulzyckii* stands out as a different species by having a much more slender, less-inflated shell and in having large sharp-edged varices.

The new Dry Tortugas species is also similar to the common *Hesperisternia multangula* (Philippi, 1848), which lives in shallow-water Turtle Grass beds all along western Florida and throughout Florida Bay (Figure 4.14I, Chapter 4). The deep-water *Hesperisternia sulzyckii* differs from its shallow-water congener in that it has a much more slender and elongated shell and in having large rounded beads on the body whorl varices. In this last character, the new species is also similar to *H. harasewychi* from the deeper-water coralline algal beds off the Dry Tortugas and western Florida (Figure 9.3C, D, Chapter 9), but differs in that it has a more slender, elongated, and less-inflated shell and in having proportionally smaller and less-developed knobs and beads on the body whorl varices. As is currently, known, five species of *Hesperisternia* occur along western Florida and the Florida Keys: *Hesperisternia grandana*, found in shallow-water sea grass beds along the northeastern Gulf of Mexico; *Hesperisternia multangula*, found in shallow-water Turtle Grass beds along western Florida and Florida Bay; *Hesperisternia sulzyckii*, found on deep-water red algae beds north of the Dry Tortugas; *Hesperisternia harasewychi*, from the even deeper-water coralline algae beds off western Florida and the Dry Tortugas; and *Hesperisternia karinae*, from the coral reef complexes of the Florida Keys and Caribbean region.

Nassariidae
Genus *Uzita* H. and A. Adams, 1853

Uzita swearingeni new species
(Figure 5.3J–L)

> **Description:** Shell of average size for genus, thin, inflated, globose, with distinctly stepped, scalariform spire; siphonal canal very short, stubby, broadly open; shoulder sharply angled, with flattened, tabulate subsutural area; body whorl and spire whorls with 14–16 thin, sharp, riblike varices; body whorl ornamented with 16–18 very thin, evenly spaced spiral cords; columella slightly recurved, with single large bladelike tooth; aperture wide and flaring, oval in shape; inner edge of lip and interior of aperture with 16 thin cords; protoconch proportionally very large, rounded, dome-like, composed of one and one-half whorls; shell color uniform canary yellow or yellow-green with lip, columella, and interior of aperture pure white; protoconch and early whorls pale orange-tan.
>
> **Holotype:** Length 7 mm, width 5 mm, USNM1231412. Paratype, length 8.5 mm, same locality as the holotype, in the research collection of the senior author.
>
> **Type Locality:** Found in a sand pocket on a sponge bioherm, 1.5-m depth off the northern end of Middle Torch Key, Lower Florida Keys, Monroe County, Florida.
>
> **Etymology:** Named for Clifford Swearingen, Fort Lauderdale, Florida, who assisted the senior author in the field while conducting research in the Florida Keys.
>
> **Discussion:** This new species of Florida Keys nassariid is most similar to the widespread western Atlantic *Uzita alba* (Say, 1826) (Figure 6.4I, J, Chapter 6), but differs in that it has a smaller shell with a much more angled shoulder; in having distinctly flattened subsutural areas and a stepped, scalariform spire; in having thinner and more numerous axial ribs (14–16 in *swearingeni* and 12–13 in *alba*); and in having fewer and finer spiral cords. The shell colors also differ greatly between the new species, with *U. alba* a cream-white with three pale brown bands and three large brown patches on the edge of the lip and with *U. swearingeni* a uniform yellow or yellow-green with a white lip. The new species is also similar to the sympatric *Uzita websteri* Petuch and Sargent, 2011 (Figure 5.3G, H, Chapter 5), which differs from *Uzita swearingeni* in that it has a larger, more globose shell with a more rounded shoulder; in having fewer axial ribs (10–12 in *websteri*); and in that it is a much more colorful shell, typically a dark yellow-tan with an orange-brown band around the sutural area and also around the base of the siphonal canal. Both *U. websteri* and *U. swearingeni* appear to be endemic to the

Florida Keys, with both species living in association with shallow-water sponge bioherms on exposed limestone seafloors.

Volutoidea
Olividae
Olivinae
Genus *Americoliva* Petuch, 2013

Americoliva matchetti new species
(Figure 9.6D–F)

> **Description:** Shell small for genus, heavy and thickened, broad and stocky, wide across shoulder, with high subpyramidal spire; shoulder distinctly angled, widest around posterior one-third of body whorl; shoulder area thickened, swollen, producing wide callous-like band that protrudes slightly from body whorl outline; sutural callous of spire whorls extremely well developed, forming characteristic ramp that extends beyond plane of suture; aperture proportionally wide, flaring; columella with 10–12 large, thin, evenly spaced plicae; fasciole proportionally small, restricted to extreme anterior end of shell; protoconch large, rounded, dome-like, composed of two whorls; shell color uniform cream-white, overlaid with scattered very faint, pale pinkish-tan zigzag flammules; edge of suture marked with numerous evenly spaced pale pinkish-tan hairlines; interior of aperture and protoconch white.
>
> **Holotype:** Length 18.5 mm, width 8 mm, USNM1231405. Paratype, length 15 mm, from the same locality and depth as the holotype, in the research collection of the senior author.
>
> **Type Locality:** Dredged from a mud-and-shell rubble bottom, 250-m depth off Miami, Dade County, Florida, along the northern end of the Miami Terrace.
>
> **Etymology:** Named for Eddie Matchett of Okeechobee, Florida, field assistant to the senior author.
>
> **Discussion:** This small deep-water olive shell is unlike any other *Americoliva* species known from Florida. With its stocky, inflated body and high, conical spire, *Americoliva matchetti* most closely resembles the widespread Caribbean Province *A. reticularis* (Lamarck, 1810), which is an abundant species in shallow intertidal sand flats in the Bahamas and Cuba. The deep terrace-dwelling *A. matchetti* differs from its wide-ranging Caribbean congener in that it is a much smaller and stockier species, with a proportionally much more truncated, less-elongated body whorl, and in having a sharply angled, swollen shoulder. *Americoliva reticularis* also lacks the protruding callous on the spire whorls and does not have the distinctly turreted spire shape that is seen in the new species. These scalariform spire characteristics of *A. matchetti* appear to be unique and have not been seen in any other western Atlantic *Americoliva* species.

Americoliva recourti new species
(Figure 9.67A–C)

> **Description:** Shell small for genus, cylindrical, without distinct shoulder, and with slightly rounded sides; spire proportionally low, subpyramidal, with spire whorls only slightly calloused; aperture narrow; columella with 10–12 large thickened plications; protoconch proportionally large, rounded, dome-like, composed of two whorls; shell color pale cream-white overlaid with two

bands of very faint, diffuse, pale pinkish-tan zigzag flammules; interior of aperture white.

Holotype: Length 23.5 mm, width 10 mm, USNM1231407. Paratypes: length 22 mm, from the same locality and depth as the holotype, in the research collection of the senior author; lengths 22 mm, 19 mm, and 16 mm, in the collection of Pierre Recourt, Egmond aan Zee, the Netherlands.

Type Locality: Dredged from a mud-and-shell rubble bottom, 250-m depth off Key Largo, Upper Florida Keys, Monroe County, Florida, along the southern end of the Miami Terrace.

Etymology: Named for Pierre Recourt of the Netherlands, one of the top workers in the Olividae and the premier authority on the *Americoliva* species of North and South America.

Discussion: In Florida, this small, cylindrically shaped olive is most similar to *Americoliva sunderlandi* (Petuch, 1986) (Figure 9.3J, K, Chapter 9), which occurs in deep water (200-m depth) off western Florida and the Dry Tortugas. Although obviously closely related, *Americoliva recourti* differs from its deep-water western Florida congener in that it has a smaller, stumpier, broader, and less-elongated and narrow shell and in having a flatter, less-protracted spire. The colors of the two species also differ greatly, with *A. recourti* white or cream-white with scattered pale pinkish-tan zigzag flammules and with *A. sunderlandi* a dark golden-tan overlaid with two, almost-solid, bands of dark brown flammules. The spire callous of *A. recourti* is pure white or pale cream-white, while that of *A. sunderlandi* is a dark brownish-tan. The new species, along with the previously described *A. matchetti*, is confined to the area of the Miami Terrace, and *A. sunderlandi* is confined to the far edge of the continental shelf off western Florida and to the Tortugas Valleys. Taking into account the shallow-water *A. sayana* and *A. bollingi*, Florida is now known to house five different species of olivids.

Bivalvia
Pteriomorphia
Arcoida
Arcidae
Arcinae
Genus *Arca* Linnaeus, 1758

Arca rachelcarsonae new species
(Figure 8.4A–C)

Description: Shell small for genus, thin, inflated, with truncated trapezoidal outline; posterior end of valves longer than anterior end; hinge line narrow, almost entire length of shell, edged with numerous fine taxodont teeth; beaks highly elevated above hinge line, sharply pointed, recurved inward; ligamental area exceptionally broad and wide, smooth and shiny, ornamented with numerous fine longitudinal lines; posterior end of valves bifurcated, with indented area between ligamental process and sloping transverse fold; byssal notch thin and narrow, covering approximately one-third of ventral area; valves sculptured with very numerous fine, equal-size concentric and radial riblets, producing cancellate, beaded texture; riblets on bifurcated posterior end covered with large scales; shell color uniform pale yellow-white, with scattered tiny brown dots on some of the riblet beads; periostracum thick, dark brown, shaggy.

Holotype: Length 16 mm, width 12 mm, USNM1231406. Paratype, length 21 mm, from the same locality and depth as the holotype, in the research collection of the senior author.

Type Locality: Found attached, by byssal threads, to a commercial lobster trap brought up from 180-m depth off Key Largo, Florida Keys Reef Tract, Upper Florida Keys, Monroe County, Florida.

Etymology: Named for Rachel Carson, renowned author, marine biologist, founder of the modern environmental protection movement in the United States, and lover of the Florida Keys.

Discussion: This unusual new deep-water arcid bivalve is only the third species of its genus to be described from the western Atlantic. The other two members of the genus *Arca* (sensu stricto), *Arca zebra* (Swainson, 1833) (Figure 5.8L, Chapter 5) and *Arca imbricata* Bruguiere, 1789 (Figure 5.8H, Chapter 5), both live in very shallow water and are abundant on reefs and rocky platforms from North Carolina south to Brazil. Of these two congeners, *A. rachelcarsonae* is most similar to *A. imbricata*, especially in having a cancellate beaded shell sculpture, but differs in that it has a much smaller and less-elongated shell; in having a bifurcated, notched posterior end; in that it is broader and much more inflated; in having a much wider ligamental shelf; and in having a much smaller and narrower byssal notch. Presently, *A. rachelcarsonae* has been recorded only from the deep-reef talus slopes off the Florida Keys Reef Tract, and it appears that this unusual arcid is endemic to the Florida Keys.

Pectinoida
Pectinoidea
Pectinidae
Pectininae

Genus *Euvola* Dall, 1897

Euvola marshallae new species
(Figure 9.4A)

Description: Shell of average size for genus, thin and fragile, nearly circular in outline; valves very flattened, bulging slightly in middle, with bottom valve slightly more convex than upper valve; outer surface of valves shiny and polished, ornamented with very low, only slightly raised radial ribs and extremely numerous fine, almost microscopic radial hairlines, producing a silky texture under magnification; interior of valves ornamented with 45–50 large, prominent radial ribs, most of which are arranged in pairs; auricles of upper valve proportionally small, symmetrical, completely adherent, rounded along edges; hinge line of upper valve almost straight, dipping only slightly in middle near umbos; auricles of lower valve slightly asymmetrical, with posterior auricle rounded and indented at base, producing slightly detached appearance; anterior auricle smaller, completely adherent, rounded on edge; hinge line of lower valve curved, distinctly concave; upper valve colored mauve and dark pinkish-tan, with variable numbers of lighter tan-pink wide radiating lines, producing sun-ray appearance; younger areas of valve overlaid with extremely numerous very small dark tan and yellow-tan dots, producing a net-like pattern; netted pattern darker and better developed on radiating lines; bottom valve pure white with dark golden-yellow band around periphery and on auricles; prodissiconch area of bottom valve colored pale tan-pink.

Holotype: Height 46 mm, width 46 mm, USNM1231408. Paratypes, heights 47 mm and 52 mm, from the same locality and depth as holotype, in the research collection of the senior author.

Type Locality: Dredged by commercial deep-water shrimpers ("Royal Red Shrimp") from 200-m depth, 20 km south of Garden Key, Dry Tortugas, Florida Keys.

Etymology: Named for Carole Marshall of West Palm Beach, Florida, the preeminent authority on the mollusks of the Lake Worth Lagoon and an avid collector of the Pectinidae.

Discussion: This common deep water scallop, which ranges throughout the Gulf of Mexico and south to northern South America, has, for over 100 years, been referred to as "*Euvola papyracea* Gabb, 1873" by most American malacologists (ie. Abbott, 1974). That species, it turns out, is actually a 20 million year old fossil from the early Miocene Baitoa Formation of the Dominican Republic and differs from the living *E. marshallae* is being a larger, more elongated, and less rounded species with proportionally larger auricles.

View, from the surface, of the extensive sponge bioherms off Middle Torch Key, Lower Florida Keys, showing a large specimen of the Vase Sponge *Ircinia campana*. (Photograph by Ron Bopp)

List of macromollusks: Florida Keys and adjacent areas

Every known macromollusk that has been reported from the marine and estuarine areas of the South Florida Bight, Florida Keys, Florida Bay, the Dry Tortugas, Biscayne Bay, Palm Beach, and the Ten Thousand Islands is arranged systematically and listed below. This biodiversity list is a compilation of personal observations from 40 years of collecting in these areas and data taken from Abbott (1974), Lyons and Quinn (1995), Mikkelsen and Bieler (2008), Petuch and Sargent (2011c), Petuch (2013), and Tucker (2013). All bathymetric ranges are taken into consideration, including the supratidal, intertidal, shallow subtidal, inner neritic, outer neritic, and bathyal zones. Only the 86 families of macrogastropods and 54 families of macrobivalves that were discussed in the Introduction are listed here. Those taxa that were listed, discussed, or illustrated in this book, but are from outside the immediate area of the Florida Keys, are given the following letter designations:

P = restricted to the deep reefs and coastal lagoons of the Palm Beach Provinciatone
T = found in the Ten Thousand Islands and Suwannean Subprovince
D = found in deeper-water areas north of the Dry Tortugas and southwestern Florida
E = species that are endemic to the Florida Keys

Class Gastropoda
Subclass Eogastropoda
Order Patellogastropoda
Suborder Nacellina
Superfamily Acmaeoidea
Family Lottiidae
    Collisella leucopleura (Gmelin, 1791)
    Lottia antillarum (Sowerby I, 1831)
    Patelloida pulcherrima (Petit, 1856)
    Patelloida pustulata (Helbling, 1779)
Subclass Orthogastropoda
Order Vetigastropoda
Superfamily Pleurotomarioidea
Family Pleurotomariidae
    Bayerotrochus midas (Bayer, 1965)
    Entemnotrochus adansonianus (Crosse and Fischer, 1861)
    Perotrochus amabilis (Bayer, 1963)

Family Haliotidae
   Haliotis pourtalesi Dall, 1881
Superfamily Fissurelloidea
Family Fissurellidae
   Diodora arcuata (Dall, 1889)
   Diodora cayenensis (Lamarck, 1822)
   Diodora dysoni (Reeve, 1850)
   Diodora fluviana Dall, 1889
   Diodora fragilis Farfante and Henriquez, 1947
   Diodora jaumei Aguayo and Rehder, 1936
   Diodora listeri (D'Orbigny, 1842)
   Diodora meta (von Ihering, 1927)
   Diodora minuta (Lamarck, 1822)
   Diodora sayi (Dall, 1899)
   Diodora viridula (Lamarck, 1822)
   Diodora wetmorei Farfante, 1945
   Diodora (Glyphis) tanneri Verrill, 1883
   Emarginula dentigera Heilprin, 1886
   Emarginula phrixoides Dall, 1927
   Emarginula pumila (A. Adams, 1851)
   Emarginula sicula Gray, 1825
   Emarginula tuberculosa Libassi, 1859
   Fissurella rosea (Gmelin, 1791)
   Fissurella (Clypidella) fascicularis Lamarck, 1822
   Fissurella (Clypidella) punctata Fischer, 1857
   Hemitoma octoradiata (Gmelin, 1791)
   Hemitoma (Montfortia) emarginata (Blainville, 1825)
   Lucapina aegis (Reeve, 1850)
   Lucapina eolis Farfante, 1945
   Lucapina philippiana (Finlay, 1930)
   Lucapina sowerbii (Sowerby, 1835)
   Lucapina suffusa (Reeve, 1850)
   Lucapinella limatula (Reeve, 1850)
   Nesta atlantica Farfante, 1947
   Puncturella acuminata Watson, 1883
   Puncturella agger Watson, 1883
   Puncturella billsae Farfante, 1947
   Puncturella erecta Dall, 1889
   Puncturella granulata Sequenza, 1863
   Rimula aequisculpta Dall, 1927
   Rimula dorriae Farfante, 1949
   Rimula frenulata Dall, 1889
   Rimula pycnonema Pilsbry, 1943
   Zeidora bigelowi Perez-Farfante, 1947
Superfamily Turbinoidea
Family Turbinidae
Subfamily Turbininae
   Turbo (Marmarostoma) castanea (Gmelin, 1791)
   Turbo (Marmarostoma) castanea crenulatus (Gmelin, 1791)

Turbo (Taenioturbo) canaliculatus Hermann, 1781
Turbo (Taenioturbo) cailletii Fischer and Bernardi, 1856

Subfamily Astraeinae
Astralium phoebia Röding, 1798
Lithopoma americana (Gmelin, 1791) **E**
Lithopoma caelata (Gmelin, 1791)
Lithopoma tuber (Linnaeus, 1767)

Family Liotiidae
Arene cruentata (Megerle von Mühlfeld, 1829)
Arene vanhyningi Rehder, 1943 **E**
Cyclostrema cancellata Marryat, 1818
Cyclostrema huesonicum Dall, 1927 **E**
Cyclostrema tortuganum (Dall, 1927) **E**
Marevalvata bairdi (Dall, 1889)
Marevalvata briareus (Dall, 1881)
Marevalvata tricarinata (Stearns, 1872)
Marevalvata trullata (Dall, 1889)
Marevalvata variabilis (Dall, 1886)
Sansonia tuberculata (Watson, 1883)

Family Phasianellidae
Eulithidium affinis (C.B. Adams, 1850)
Eulithidium bella (M. Smith, 1937)
Eulithidium plerocladica (Robertson, 1958) **D**
Eulithidium thalassicola (Robertson, 1958)

Superfamily Trochoidea
Family Trochidae
Subfamily Tegulinae
Tegula (Agathistoma) excavata (Lamarck, 1822)
Tegula (Agathistoma) fasciata (Born, 1778)
Tegula (Agathistoma) hotessieriana (d'Orbigny, 1842)
Tegula (Agathistoma) lividomaculata (C.B. Adams, 1845)

Subfamily Stomatellinae
Pseudostomatella erythrocoma (Dall, 1899)
Synaptocochlea nigrita Rehder, 1939 **E**
Synaptocochlea picta (d'Orbigny, 1842)

Family Margaritidae
Gaza superba (Dall, 1881)
Microgaza inornata Dall, 1881
Microgaza rotella Dall, 1881

Family Solariellidae
Solariella aegleis Watson, 1879
Solariella lacunella (Dall, 1881)
Solariella lamellosa Verrill and Smith, 1880
Solariella lissocoma (Dall, 1881)
Solariella periscopia (Dall, 1927)
Solariella pourtalesi Clench and Aguayo, 1939
Suavotrochus lubricus (Dall, 1881)

Family Calliostomatidae
Calliostoma adelae Schwengel, 1951 **E**

Calliostoma euglyptum (A. Adams, 1846)
Calliostoma fascinans Schwengel and McGinty, 1942
Calliostoma frumari Garcia, 2007 **E**
Calliostoma javanicum (Gmelin, 1791)
Calliostoma jujubinum (Gmelin, 1791)
Calliostoma marionae Dall, 1906
Calliostoma pulchrum (C.B. Adams, 1850)
Calliostoma roseolum Dall, 1880
Calliostoma sapidum Dall, 1881
Calliostoma scalenum Quinn, 1992
Calliostoma tampaense (Conrad, 1846)
Calliostoma yucatecanum Dall, 1881
Calliostoma (Kombologion) bairdii Verrill and Smith, 1880
Calliostoma (Kombologion) benedicti Dall, 1889
Calliostoma (Kombologion) hendersoni Dall, 1927
Calliostoma (Kombologion) oregon Clench and Turner, 1960
Calliostoma (Kombologion) psyche Dall, 1889
Calliostoma (Kombologion) sayanum Dall, 1889

Order Neritopsina
Superfamily Neritoidea
Family Neritidae
Nerita (Ritena) versicolor Gmelin, 1791
Nerita (Linnerita) peloronta Linnaeus, 1758
Nerita (Theliostyla) fulgurans Gmelin, 1791
Nerita (Theliostyla) lindae Petuch, 1988 **E**
Nerita (Theliostyla) tessellata Gmelin, 1791
Neritina (Vitta) clenchi Russel, 1940
Neritina (Vitta) usnea (Röding, 1798)
Neritina (Vitta) virginea (Linnaeus, 1758)
Puperita pupa (Linnaeus, 1758)
Smaragdia viridemaris Maury, 1917

Family Phenacolepadidae
Phenacolepas hamillei (Fischer, 1857)

Superorder Caenogastropoda
Order Sorbeoconcha
Superfamily Cerithioidea
Family Cerithiidae
Bayericerithium litteratum (Born, 1778)
Bayericerithium semiferrugineum (Lamarck, 1822)
Bittiolum varium (Pfeiffer, 1840)
Cerithium atratum (Born, 1778)
Cerithium chara Pilsbry, 1949 **D**
Cerithium eburneum Bruguiere, 1792
Cerithium guinaicum Philippi, 1849
Cerithium lindae Petuch, 1987 **P**
Cerithium lutosum Menke, 1828
Cerithium lymani Pilsbry, 1949 **D**
Cerithium muscarum Say, 1822

Family Batillariidae
  Batillaria minima (Gmelin, 1791)
Family Potamididae
  Cerithideopsis costatus (da Costa, 1778)
  Cerithideopsis scalariformis (Say, 1825)
Family Planaxidae
  Angiola lineata (da Costa, 1778)
  Planaxis (Supplanaxis) nucleus (Bruguiere, 1789)
Family Modulidae
  Modulus calusa Petuch, 1988 **E**
  Modulus calusa foxhalli Petuch and Myers, new subspecies **P**
  Modulus floridanus Conrad, 1869 **T**
  Modulus kaicherae Petuch, 1987
  Modulus lindae Petuch, 1987
  Modulus pacei Petuch, 1987 **P**
Family Turritellidae
  Torcula exoleta (Linnaeus, 1758)
  Torculoidella acropora (Dall, 1889)
  Torculoidella lindae Petuch, 1987 **D**
  Vermicularia fargoi owensi Petuch and Myers, new subspecies **T**
  Vermicularia knorri (Deshayes, 1843)
  Vermicularia spirata (Philippi, 1836)
Family Siliquariidae
  Siliquaria modesta Dall, 1881
  Siliquaria squamata Blainville, 1827
Suborder Hypsogastropoda
Infraorder Littorinimorpha
Superfamily Littorinoidea
Family Littorinidae
  Amerilittorina angustior (Mörch, 1876)
  Amerilittorina jamaicensis (C.B. Adams, 1850)
  Amerilittorina placida (Reid, 2009)
  Amerilittorina ziczac (Gmelin, 1791)
  Cenchritis muricatus (Linnaeus, 1758)
  Echinolittorina antonii (Philippi, 1846)
  Echinolittorina dilatata (d'Orbigny, 1841)
  Echinolittorina tuberculata (Menke, 1828)
  Fossarilittorina meleagris (Potiez and Michaud, 1838)
  Littoraria angulifera (Lamarck, 1822)
  Littoraria nebulosa (Lamarck, 1822) (and color form *tessellata* Philippi, 1847)
  Melarhaphe mespillum (Megerle von Mühlfeld)
  Tectininus nodulosus (Pfeiffer, 1839)
Superfamily Rissooidea
Family Truncatellidae
  Truncatella caribaeensis Reeve, 1842
  Truncatella pulchella Pfeiffer, 1839
Superfamily Stromboidea
Family Strombidae
  Alliger gallus (Linnaeus, 1758)

Eustrombus gigas (Linnaeus, 1758)
Lobatus raninus (Gmelin, 1791)
Macrostrombus costatus (Gmelin, 1791)
Strombus alatus Gmelin, 1791
Strombus pugilis Linnaeus, 1758

Superfamily Vanikoroidea
Family Vanikoridae
Vanikoro oxychone Mörch, 1877

Family Hipponicidae
Hipponix antiquatus (Linnaeus, 1767)
Hipponix subrufus (Lamarck, 1822)
Malluvium benthophilum (Dall, 1889)

Superfamily Calyptraeoidea
Family Calyptraeidae
Bostrycapulus aculeatus (Gmelin, 1791)
Calyptraea centralis (Conrad, 1841)
Cheilea equestris (Linnaeus, 178)
Crepidula convexa Say, 1822
Crepidula fornicata (Linnaeus, 1758) **P, T**
Crepidula maculosa Conrad, 1846 **T**
Crepidula ustulatulina (Collin, 2002)
Crucibulum auricula (Gmelin, 1791)
Crucibulum striatum Say, 1824
Ianacus atrasolea (Collin, 2002)
Ianacus plana (Say, 1822) **T, P**

Superfamily Capuloidea
Family Capulidae
Capulus ungaricus (Linnaeus, 1767)
Capulus (Krebsia) intortus (Lamarck, 1822)

Superfamily Xenophoroidea
Family Xenophoridae
Trochotugurium longleyi (Bartsch, 1931)
Tugurium caribbaeum (Petit, 1856)
Xenophora conchyliophora (Born, 1780)
Xenophora microdiscus Petuch, 1994 (named as a late Pleistocene fossil but still extant) **D**

Superfamily Vermetoidea
Family Vermetidae
Dendropoma corrodens (d'Orbigny, 1841) (= *annulatus* Daudin)
Dendropoma irregulare (d'Orbigny, 1841)
Petaloconchus erectus (Dall, 1888)
Petaloconchus mcgintyi (Olsson and Harbison, 1953)
Petaloconchus varians (d'Orbigny, 1841)
Serpulorbis decussatus (Gmelin, 1791)
Vermetus (Thylaeodus) nigricans (Dall, 1884) **T**

Superfamily Cypraeoidea
Family Cypraeidae
Erosaria acicularis (Gmelin, 1791)
Luria cinerea (Gmelin, 1791)
Macrocypraea zebra (Linnaeus, 1758)

Macrocypraea (Lorenzicypraea) cervus (Linnaeus, 1771)
Propustularia surinamensis (Perry, 1811)
Family Ovulidae
Subfamily Simniinae
Adamantia solemi Cate, 1973 **E**
Calcarovula piragua (Dall, 1889)
Cymbovula acicularis (Lamarck, 1810)
Cyphoma alleneae Cate, 1973 **E** (may also be found in Cuba)
Cyphoma gibbosum (Linnaeus, 1758)
Cyphoma lindae Petuch, 1987 **D**
Cyphoma mcgintyi Pilsbry, 1939
Cyphoma rhomba Cate, 1978 **E**
Cyphoma robustior Bayer, 1941 **D**
Cyphoma sedlaki Cate, 1976 **E**
Cyphoma signatum Pilsbry and McGinty, 1939
Pseudocyphoma aureocincta (Dall, 1889)
Pseudocyphoma gibbulum (Cate, 1978) **E**
Pseudocyphoma intermedium (Sowerby II, 1828)
Pseudosimnia sphoni Cate, 1973 **E**
Pseudosimnia vanhyningi (M. Smith, 1940)
Simniulena uniplicata (Sowerby II, 1849)
Family Pediculariidae
Pedicularia decussata (Gould, 1855)
Superfamily Velutinoidea
Family Lamellariidae
Lamellaria leucosphaera Schwengel, 1942
Lamellaria perspicua (Linnaeus, 1758)
Family Triviidae
Cleotrivia candidula (Gaskoin, 1836)
Dolichupis leei Fehse and Grego, 2002
Dolichupis leucosphaera Schilder, 1931
Hespererato maugeriae (Gray, 1832)
Niveria antillarum (Schilder, 1922)
Niveria maltbiana (Schwengel and McGinty, 1942)
Niveria nix (Schilder, 1922) (Figure 8.3K)
Niveria quadripunctata (Gray, 1827)
Pusula juyingae Petuch and Sargent, 2011 **P** (incorrectly referred to as *P. costispunctata*)
Pusula pediculus (Linnaeus, 1758)
Pusula pullata (Sowerby II, 1870)
Pusula suffusa (Gray, 1832)
Superfamily Naticoidea
Family Naticidae
Haliotinella patinaria (Guppy, 1876)
Lunatia fringilla (Dall, 1881)
Lunatia perla (Dall, 1881)
Natica castrensis Dall, 1889
Natica livida Pfeiffer, 1840
Natica marochiensis (Gmelin, 1791)
Natica perlineata Dall, 1889

Naticarius canrena (Linnaeus, 1758)
Naticarius tedbayeri (Rehder, 1986)
Neverita delessertiana (Recluz, 1843) **T** (not found in the Florida Keys)
Neverita delessertiana patriceae Petuch and Myers, new subspecies **P**
Neverita nubila (Dall, 1925)
Polinices hepaticus (Röding, 1798)
Polinices lacteus (Guilding, 1834)
Polinices uberinus (d'Orbigny, 1842)
Sigatica carolinensis (Dall, 1889)
Sigatica semisulcata (Gray, 1839)
Sinum maculosum (Say, 1831)
Sinum minor (Dall, 1889)
Sinum perspectivum (Say, 1831)
Stigmaulax sulcatus (Born, 1778)
Tectonatica pusilla (Say, 1822)

Superfamily Tonnoidea
Family Tonnidae
Tonna galea (Linnaeus, 1758)
Tonna pennata (Mörch, 1852)

Family Ficidae
Ficus carolae Clench, 1945
Ficus communis Röding, 1798 **T** (not found in the Florida Keys)

Family Cassidae
Subfamily Cassinae
Casmaria atlantica Clench, 1944
Cassis flammea (Linnaeus, 1758)
Cassis madagascariensis Lamarck, 1822
Cassis spinella Clench, 1944
Cassis tuberosa (Linnaeus, 1758)
Cypraecassis testiculus (Linnaeus, 1758)
Echinophoria coronadoi (Crosse, 1867)
Sconsia striata (Lamarck, 1816) (incorrectly called *S. grayi* Adams, 1855 by some workers)
Semicassis cicatricosum (Gmelin, 1791)
Semicassis granulatum (Born, 1778)

Subfamily Oocorythinae
Benthodolium abyssorum (Verrill and Smith, 1884)
Eudolium crosseanum (Monterosato, 1869)
Eudolium thompsoni McGinty, 1955 **E** (often incorrectly synonymized with *crosseanum*)
Oocorys bartschi Rehder, 1943
Oocorys sulcata Fischer, 1883

Family Bursidae
Bufonaria bufo (Bruguiere, 1792)
Bursa cubaniana (d'Orbigny, 1842)
Bursa (Colubrellina) corrugata (Perry, 1811)
Bursa (Colubrellina) ranelloides tenuisculpta Dautzenberg and Fischer, 1906
Bursa (Lampasopsis) grayana Dunker, 1862
Bursa (Lampasopsis) rhodostoma thomae (d'Orbigny, 1842)

Family Ranellidae
Charonia variegata (Lamarck, 1816)

Cymatium femorale (Linnaeus, 1758)
Cymatium (Gutturnium) muricinum (Röding, 1798)
Cymatium (Gutturnium) nicobaricum (Röding, 1798)
Cymatium (Linatella) cingulatum (Lamarck, 1822)
Cymatium (Monoplex) aquatile (Reeve, 1844)
Cymatium (Monoplex) mundum (Gould, 1849)
Cymatium (Monoplex) parthenopeum (von Salis, 1793)
Cymatium (Monoplex) pileare (Linnaeus, 1758)
Cymatium (Ranularia) cynocephalum (Lamarck, 1816) (= caribbaeum) Clench and Turner
Cymatium (Ranularia) rehderi Verrill, 1950
Cymatium (Septa) krebsii Mörch, 1877
Cymatium (Septa) pharcidum Dall, 1889
Cymatium (Septa) vespaceum (Lamarck, 1822)
Cymatium (Tritoniscus) labiosum (Wood, 1828)

Family Personiidae

Distorsio clathrata (Lamarck, 1816)
Distorsio mcgintyi Emerson and Puffer, 1953
Distorsio perdistorta Fulton, 1938

Superfamily Janthinoidea

Family Janthinidae

Janthina janthina (Linnaeus, 1758)
Janthina (Jodina) exigua Lamarck, 1816
Janthina (Violetta) globosa Swainson, 1822
Janthina (Violetta) pallida (Thompson, 1840)
Recluzia rollandiana Petit, 1853

Family Epitoniidae

Alexania floridana (Pilsbry, 1945)
Amaea mitchelli (Dall, 1896)
Amaea retifera (Dall, 1889)
Asperiscala apiculata (Dall, 1889)
Asperiscala babylonium (Dall, 1889)
Asperiscala candeanum (d'Orbigny, 1842)
Asperiscala denticulata (Sowerby I, 1844)
Asperiscala frielei (Dall, 1889)
Asperiscala fractum (Dall, 1927)
Asperiscala multistriata (Say, 1826)
Asperiscala novangliae (Couthouy, 1838)
Asperiscala polacia (Dall, 1889)
Asperiscala pourtalesi (Verrill and Bush, 1880)
Asperiscala rushii (Dall, 1889)
Boreoscala blainei Clench and Turner, 1953
Cirsotrema dalli Rehder, 1945
Cirsotrema pilsbryi (McGinty, 1940)
Cycloscala echinaticosta (d'Orbigny, 1842)
Cylindriscala andrewsi (Verrill, 1882)
Cylindriscala watsoni (deBoury, 1911)
Dentiscala burryi Clench and Turner, 1950
Dentiscala crenata (Linnaeus, 1758)
Dentiscala hotessieriana (d'Orbigny, 1842)

Depressiscala nautlae (Mörch, 1874)
Depressiscala nitidella (Dall, 1889)
Epitonium albidum (Say, 1830)
Epitonium angulatum (Say, 1830)
Epitonium championi Clench and Turner, 1952
Epitonium foliaceicostum (d'Orbigny, 1842)
Epitonium humphreysi (Kiener, 1838)
Epitonium krebsii (Mörch, 1874)
Epitonium occidentale (Nyst, 1871)
Epitonium tollini Bartsch, 1938
Epitonium unifasciatum (Sowerby I, 1844)
Gyroscala lamellosa (Lamarck, 1822)
Nodiscala aurifilia (Dall, 1889)
Nodiscala eolis Clench and Turner, 1950
Nodiscala pumilio (Mörch, 1874)
Opalia abbotti Clench and Turner, 1952
Opaliopsis atlantis Clench and Turner, 1952
Opaliopsis cania (Dall, 1927)
Opaliopsis concava (Dall, 1889) **E**
Opaliopsis nitida (Verrill and Smith, 1885)
Opaliopsis opalina (Dall, 1927)
Pictoscala rupicola (Kurtz, 1860)
Stenorhytis pernobilis (Fischer and Bernardi, 1857)
Superfamily Eulimoidea
Family Eulimidae (only species with large shells are listed here)
Niso aeglees Bush, 1885
Niso hendersoni Bartsch, 1953
Infraorder Neogastropoda
Superfamily Muricoidea
Family Muricidae
Subfamily Muricinae
Aspella anceps (Lamarck, 1822)
Aspella castor Radwin and D'Attilio, 1976
Attiliosa philippiana (Dall, 1889)
Calotrophon ostrearum (Conrad, 1846)
Chicoreus dilectus (A. Adams, 1855)
Chicoreus mergus E. Vokes, 1974
Chicoreus rachelcarsonae Petuch, 1987 **D**
Dermomurex pacei Petuch, 1988
Dermomurex pauperculus (C.B. Adams, 1856)
Dermomurex (Gracilimurex) elizabethae McGinty, 1940
Dermomurex (Trialatella) glicksteini Petuch, 1987 **P**
Hexaplex fulvescens (Sowerby I, 1834)
Paziella pazi (Crosse, 1869)
Phyllonotus oculatus (Reeve, 1845)
Phyllonotus pomum (Gmelin, 1791)
Phyllonotus whymani Petuch and Sargent, 2011 **D**
Poirieria nuttingi (Dall, 1896)
Pterochelus ariomus (Clench and Farfante, 1945)

Pteropurpura bequaerti (Clench and Farfante, 1945)
Pterynotus bushae E. Vokes, 1970
Pterynotus guesti Harasewych and Jensen, 1979 **E**
Pterynotus phaneus (Dall, 1889)
Pterynotus tristichus (Dall, 1889)
Siratus articulatus (Reeve, 1845)
Siratus beaui (Fischer and Bernardi, 1857)
Siratus consuela (Verrill, 1950)
Vokesimurex cabritti (Bernardi, 1859)
Vokesimurex lindajoyceae (Petuch, 1987) **D**
Vokesimurex morrisoni Petuch and Sargent, 2011
Vokesimurex rubidus (F. C. Baker, 1897)
Vokesimurex rubidus form *marcoensis* (Sowerby III, 1900) **E** (may be a distinct species)
Vokesimurex tryoni (Hidalgo, 1880)

Subfamily Muricopsinae
Caribiella alveata (Kiener, 1842)
Favartia cellulosa (Conrad, 1846)
Favartia emipowlusi (Abbott, 1954)
Favartia goldbergi Petuch and Sargent, 2011 **P**
Favartia lindae Petuch, 1987 **D**
Favartia minirosea (Abbott, 1954)
Favartia pacei Petuch, 1988 **E**
Murexiella caitlinae Petuch and Myers, new species **E**
Murexiella glypta (M. Smith, 1938)
Murexiella hidalgoi (Crosse, 1869)
Murexiella kalafuti Petuch, 1987 **E**
Murexiella levicula (Dall, 1889)
Murexiella taylorae Petuch, 1987 **D**
Muricopsis oxytatus (M. Smith, 1938)
Pygmaepterys richardbinghami (Petuch, 1987) **P**
Risomurex caribbaeus (Bartsch and Rehder, 1939)
Risomurex deformis (Reeve, 1846)
Risomurex muricoides (C.B. Adams, 1845)
Risomurex roseus (Reeve, 1846)

Subfamily Ergalitaxinae
Orania grayi (Dall, 1889)
Trachypollia nodulosa (C.B. Adams, 1845)
Trachypollia sclera Woodring, 1928
Trachypollia turricula (Maltzan, 1884)

Subfamily Ocenebrinae
Eupleura sulcidentata Dall, 1890
Trossulasalpinx macra (Verrill, 1887)

Subfamily Rapaninae
Mancinella deltoidea (Lamarck, 1822)
Plicopurpura patula (Linnaeus, 1758)
Stramonita buchecki Petuch, 2013 **P**
Stramonita floridana (Conrad, 1837)
Stramonita rustica (Lamarck, 1822)

Subfamily Typhinae
Pterotyphis pinnatus (Broderip, 1833)
Tripterotyphis triangularis (A. Adams, 1856)
Typhinellus sowerbii (Broderip, 1833)
Subfamily Magilinae (sometimes placed in the separate family Coralliophilidae)
Babelomurex dalli (Emerson and Puffer, 1963)
Babelomurex fax (Bayer, 1971)
Babelomurex scalariformis (Lamarck, 1822)
Coralliophila abberans (C.B. Adams, 1850)
Coralliophila aedonis (Watson, 1886)
Coralliophila caribbaea Abbott, 1958
Coralliophila galea (Dillwyn, 1823)
Coralliophila pacei Petuch, 1987 **P**
Coralliophila richardi (Fischer, 1882) (= *lactuca*)
Quoyula kalafuti Petuch, 1987 **E**
Superfamily Buccinoidea
Family Buccinidae
Subfamily Buccininae
Agassitula agassizi (Clench and Aguayo, 1941)
Antillophos beaui (Fischer and Bernardi, 1857)
Antillophos candeanus (d'Orbigny, 1842)
Antillophos virginiae (Schwengel, 1942)
Bartschia significans Rehder, 1943 (= *frumari*)
Belomitra pourtalesi (Dall, 1881)
Colus rushii (Dall, 1889)
Eosipho canetae (Clench and Aguayo, 1944)
Liomesus stimpsoni Dall, 1889
Manaria burkeae Garcia, 2008
Parviphos adelus (Schwengel, 1942)
Ptychosalpinx globulus (Dall, 1889)
Retimohnia carolinensis (Verrill, 1884)
Subfamily Pisaniinae
Anna florida Garcia, 2008
Bailya intricata (Dall, 1884)
Bailya parva (C.B. Adams, 1850)
Bailya weberi (Watters, 1983)
Engina corinnae Crovo, 1971
Engina turbinella (Kiener, 1835)
Gemophos auritulus (Link, 1807)
Gemophos tinctus Conrad, 1846
Gemophos tinctus pacei Petuch and Sargent, 2011 **T**
Hesperisternia harasewychi (Petuch, 1987) **D**
Hesperisternia karinae (Usticke, 2008)
Hesperisternia multangulus (Philippi, 1848)
Hesperisternia sulzyckii Petuch and Myers, new species **D**
Pisania pusio (Linnaeus, 1758)
Family Colubrariidae
Colubraria testacea (Mörch, 1852)

Family Melongenidae
Melongena (Rexmela) bicolor (Say, 1827)
Melongena (Rexmela) corona (Gmelin, 1791) **T**
Melongena (Rexmela) corona winnerae Petuch, 2004 **P**
Family Busyconidae
Subfamily Busyconinae
Lindafulgur lyonsi (Petuch, 1987) **D**
Sinistrofulgur sinistrum (Hollister, 1958)
Subfamily Busycotypinae
Fulguropsis plagosum (Conrad, 1863) **D**
Fulguropsis pyruloides (Say, 1822) **T, P**
Fulguropsis spiratum keysensis Petuch, 2013 **E**
Family Fasciolariidae
Subfamily Fasciolariinae
Cinctura hunteria (Perry, 1811)
Cinctura tortugana (Hollister, 1957) **D**
Fasciolaria bullisi Lyons, 1972 **D**
Fasciolaria tulipa (Linnaeus, 1758)
Triplofusus papillosus (Sowerby I, 1825)
Subfamily Peristerniinae
Dolicholatirus cayohuesonicus (Sowerby II, 1878)
Dolicholatirus pauli (McGinty, 1955)
Leucozonia jacarusoi Petuch, 1987 **D**
Leucozonia lineata Usticke, 1969
Leucozonia nassa (Gmelin, 1791)
Leucozonia ocellata (Gmelin, 1791)
Polygona angulata (Röding, 1798)
Polygona carinifera (Lamarck, 1822)
Polygona infundibula (Gmelin, 1791)
Polygona nemata (Woodring, 1928) (= *P. cymatius* Schwengel)
Subfamily Fusininae
Fusinus excavatus (Sowerby II, 1880) (*F. eucosmius* is a synonym)
Fusinus halistreptus (Dall, 1889)
Fusinus schrammi (Crosse, 1865)
Fusinus stegeri Lyons, 1978 **D**
Harasewychia aepynota (Dall, 1889)
Harasewychia alcimus (Dall, 1889)
Harasewychia amianta (Dall, 1889)
Harasewychia amphiurgus (Dall, 1889)
Harasewychia benthalis (Dall, 1889)
Harasewychia rushii (Dall, 1889)
Heilprinia lindae Petuch, 1987 **D**
Heilprinia timessus (Dall, 1889)
Family Nassariidae
Phrontis polygonatus (Lamarck, 1822)
Phrontis vibex (Say, 1822)
Uzita alba (Say, 1826)

Uzita antillarum (d'Orbigny, 1842)
Uzita consensa (Ravenel, 1861)
Uzita hotessieri (d'Orbigny, 1845)
Uzita paucicostata (Marrat, 1877)
Uzita scissurata (Dall, 1889)
Uzita swearingeni Petuch and Myers, new species **E**
Uzita websteri Petuch and Sargent, 2011 **E**

Family Columbellidae

Aesopus stearnsi (Tryon, 1883)
Astyris lunata (Say, 1826)
Astyris multilineata (Dall, 1889)
Columbella mercatoria (Linnaeus, 1758)
Columbella rusticoides Heilprin, 1886
Columbellopsis nycteus (Duclos, 1846)
Cosmioconcha rikae Monsecour and Monsecour, 2006
Costoanachis avara (Say, 1822)
Costoanachis floridana Rehder, 1939
Costoanachis semiplicata Stearns, 1873
Costoanachis sertulariarum (d'Orbigny, 1839)
Costoanachis sparsa (Reeve, 1859)
Costoanachis translirata (Ravenel, 1861)
Metulella columbellata Dall, 1889
Mitrella argus d'Orbigny, 1842
Mitrella ocellata (Gmelin, 1791)
Nassarina bushiae (Dall, 1889)
Nassarina monilifera (Sowerby II, 1844)
Nassarina sparsipunctata (Rehder, 1943)
Nitidella laevigata (Linnaeus, 1758)
Nitidella nitida (Lamarck, 1822)
Suturoglypta hotessieriana (d'Orbigny, 1842)
Suturoglypta iontha (Ravenel, 1861)
Zafrona idalina (Duclos, 1840)
Zafrona pulchella (Blainville, 1829)
Zafrona taylorae Petuch, 1987 **E**

Superfamily Volutoidea

Family Volutidae

Subfamily Lyriinae

Enaeta cylleniformis (Sowerby I, 1844) (This small volute, which is common in the Bahamas, has been reported from northern Biscayne Bay by several collectors and workers; its presence throughout the Florida Keys reef tracts has yet to be established.)

Subfamily Scaphellinae

Aurinia dubia (Broderip, 1827) **D**
Caricellopsis matchetti (Petuch and Sargent, 2011) **E**
Clenchina dohrni (Sowerby III, 1903) **E**
Clenchina florida (Clench and Aguayo, 1940) **E**
Clenchina gouldiana (Dall, 1887)
Clenchina marionae (Pilsbry and Olsson, 1953) **D**
Rehderia georgiana (Clench, 1946)
Rehderia schmitti (Bartsch, 1931) **E**

Scaphella junonia (Lamarck, 1804) **P** (North Carolina to Palm Beach; Gulf of Mexico)

Scaphella junonia elizabethae Petuch and Sargent, 2011 **E** (shallow-water Florida Keys)

Family Olividae

Subfamily Olivinae

Americoliva bollingi (Clench, 1943) (Florida Keys and eastern Florida)

Americoliva matchetti Petuch and Myers, new species **E**

Americoliva recourti Petuch and Myers, new species **E**

Americoliva sayana (Ravenel, 1834) **T, P**

Americoliva sunderlandi (Petuch, 1987) **D**

Family Olivellidae

Callianax thompsoni (Olsson, 1956)

Jaspidella blanesi (Ford, 1898)

Jaspidella jaspidea (Gmelin, 1791)

Jaspidella miris Olsson, 1956

Macgintiella rotunda (Dall, 1889)

Macgintiella watermani (McGinty, 1940)

Niteoliva moorei (Abbott, 1951)

Olivella elongata Marrat, 1871 **P** (common in Lake Worth and off Palm Beach)

Olivella floralia (Duclos, 1853)

Olivella lactea Marrat, 1871 (= *adelae* Olsson, 1956)

Olivella mcgintyi Olsson, 1956

Olivella nivea (Gmelin, 1791)

Olivella (Dactylidia) dealbata (Reeve, 1850)

Olivella (Dactylidia) mutica (Say, 1822)

Olivella (Dactylidia) pusilla (Marrat, 1871)

Olivina bullula (Reeve, 1850)

Family Turbinellidae

Subfamily Vasinae

Vasum muricatum (Born, 1778)

Subfamily Columbariinae

Peristarium aurora (Bayer, 1971)

Peristarium electra (Bayer, 1971)

Peristarium merope (Bayer, 1971)

Family Harpidae

Subfamily Moruminae

Cancellomorum dennisoni (Reeve, 1842)

Morum oniscus (Linnaeus, 1758)

Morum purpureum Röding, 1798

Family Mitridae

Dibaphimitra florida (Gould, 1856) **E**

Mitra (Cancilla) antillensis Dall, 1889

Nebularia barbadensis (Gmelin, 1791)

Scabricola nodulosa (Gmelin, 1791)

Scabricola pallida (Usticke, 1959)

Subcancilla straminea (A. Adams, 1853)

Family Costellariidae

Nodicostellaria laterculata (Sowerby II, 1874)

Pusia puella (Reeve, 1845)

Thala foveata (Sowerby I, 1874)
Vexillum arestum (Rehder, 1943)
Vexillum dermestinum (Lamarck, 1811)
Vexillum epiphanea Rehder, 1943
Vexillum exiguum (C.B. Adams, 1845) (= *hanleyi*)
Vexillum hendersoni (Dall, 1927)
Vexillum histrio (Reeve, 1844)
Vexillum moisei McGinty, 1955
Vexillum moniliferum (C.B. Adams, 1850) (= "*albocinctum*")
Vexillum pulchellum (Reeve, 1844)
Vexillum styliolum (Dall, 1927)
Vexillum styria Dall, 1889
Vexillum sykesi (Melvill, 1925)
Vexillum variatum (Reeve, 1845)
Vexillum wandoense (Holmes, 1860)
Family Volutomitridae
Microvoluta blakeana (Dall, 1889)
Microvoluta laevior (Dall, 1889)
Family Marginellidae (see Lipe, 1991, for illustrations of the Florida marginellids)
Dentimargo aureocincta Stearns, 1872
Dentimargo eburneola Conrad, 1834
Dentimargo idiochila Schwengel, 1943 **E**
Dentimargo reducta (Bavay, 1922)
Eratoidea hematita (Kiener, 1834) **D**
Eratoidea watsoni (Dall, 1881)
Hyalina avenacea (Deshayes, 1844)
Hyalina pallida (Linnaeus, 1758)
Prunum apicinum (Menke, 1828) **T, P**
Prunum apicinum virgineum (Jousseaume, 1875) **E**
Prunum bellum (Conrad, 1868)
Prunum carneum (Storer, 1837)
Prunum cassis (Dall, 1889)
Prunum evelynae (Bayer, 1943) **P**
Prunum frumari Petuch and Sargent, 2011 **E**
Prunum guttatum (Dillwyn, 1817)
Prunum guttatum nobilianum (Bayer, 1943) **P**
Prunum hartleyana (Schwengel, 1941) **D**
Prunum redfieldi (Tryon, 1882) **E**
Prunum torticulum (Dall, 1881)
Prunum succinea (Conrad, 1846) (= *veliei* Pilsbry)
Prunum virginianum (Conrad, 1868)
Volvarina albolineata (d'Orbigny, 1842)
Volvarina avena (Kiener, 1834)
Volvarina garycooverti Espinosa and Ortea, 1998 **E**
Volvarina lactea (Kiener, 1841)
Volvarina pallida (Linnaeus, 1758)
Volvarina styria Dall, 1889
Volvarina subtriplicata (d'Orbigny, 1842)

Family Cystiscidae
Subfamily Cystiscinae
    Cysticus microgonia (Dall, 1927)
Subfamily Persiculinae
    Canalispira kerni Garcia, 2007
    Gibberula catenata (Montagu, 1803)
    Gibberula lavalleeana (d'Orbigny, 1842)
    Persicula pulcherrima (Gaskoin, 1849)
Subfamily Granulininae
    Granulina hadria (Dall, 1889)
    Pugnus serrei (Bavay, 1911)
Superfamily Cancellarioidea
Family Cancellariidae
Subfamily Cancellariinae
    Axelella agassizii (Dall, 1889)
    Bivetopsia rugosum (Lamarck, 1822)
    Cancellaria adelae Pilsbry, 1940 **E**
    Cancellaria reticulata (Linnaeus, 1758) **T, P**
    Cancellaria richardpetiti Petuch, 1987 **D**
    Ventrilia tenerum (Philippi, 1848)
Subfamily Plesiotritoninae
    Tritonoharpa janowskyi Petuch and Sargent, 2011 **P**
    Tritonoharpa lanceolata (Menke, 1828)
Superfamily Conoidea
Family Conidae
Subfamily Coninae
    Chelyconus ermineus (Born, 1778)
Subfamily Puncticulinae
    Atlanticonus granulatus (Linnaeus, 1758)
    Attenuiconus attenuatus (Reeve, 1844)
    Conasprelloides cancellatus (Hwass, 1792)
    Conasprelloides stimpsoni (Dall, 1902) (= *levistimpsoni* Tucker)
    Conasprelloides villepini (Fischer and Bernardi, 1857)
    Dauciconus amphiurgus (Dall, 1889)
    Dauciconus aureonimbosus (Petuch, 1987) **D**
    Dauciconus daucus (Hwass, 1792)
    Dauciconus glicksteini (Petuch, 1987) **P**
    Gladioconus mus (Hwass, 1792)
    Gradiconus anabathrum (Crosse, 1865) **T**
    Gradiconus anabathrum tranthami (Petuch, 1995) **E**
    Gradiconus burryae (Clench, 1942) **E**
    Gradiconus mazzolii Petuch and Sargent, 2011 **E**
    Gradiconus patglicksteinae (Petuch, 1987) **P**
    Gradiconus philippii (Kiener, 1845)
    Kellyconus binghamae (Petuch, 1987) **E**
    Kellyconus patae (Abbott, 1971)
    Lindaconus atlanticus (Clench, 1942)
    Lindaconus aureofasciatus (Rehder and Abbott, 1951) **D**

Stephanoconus regius (Gmelin, 1791)

Tuckericonus flamingo (Petuch, 1980) **P**

Tuckericonus flavescens caribbaeus (Clench, 1942)

Family Conilithidae

Subfamily Conilithinae

Dalliconus mcgintyi (Pilsbry, 1955)

Dalliconus rainseae (McGinty, 1953)

Jaspidiconus fluviamaris Petuch and Sargent, 2011 **E**

Jaspidiconus mindanus (Hwass, 1792)

Jaspidiconus pealii (Green, 1830)

Jaspidiconus pfluegeri Petuch, 2004

Jaspidiconus stearnsi (Conrad, 1869) **T**

Jaspidiconus vanhyningi (Rehder, 1944) **E**

Kohniconus delessertii (Recluz, 1843)

Family Terebridae

Fusoterebra benthalis (Dall, 1899)

Hastula cinerea (Born, 1778)

Hastula hastata (Gmelin, 1791)

Hastula maryleeae Burch, 1965

Hastula salleana (Deshayes, 1859)

Myurella taurina (Lightfoot, 1786)

Myurellina floridana (Dall, 1889)

Myurellina lindae (Petuch, 1987) **D**

Strioterebrum arcas (Abbott, 1954) **D**

Strioterebrum concavum (Say, 1827)

Strioterebrum dislocatum (Say, 1822) **T**

Strioterebrum dislocatum onslowensis (Petuch, 1974)

Strioterebrum glossema (Schwengel, 1940)

Strioterebrum limatulum (Dall, 1889)

Strioterebrum lutescens (Dall, 1889)

Strioterebrum protextum (Conrad, 1845)

Strioterebrum rushii (Dall, 1889)

Strioterebrum vinosum (Dall, 1889)

Superfamily Turroidea

Family Turridae

Gemmula periscelida (Dall, 1889)

Polystira albida (Perry, 1811)

Polystira florencae Bartsch, 1934

Polystira sunderlandi Petuch, 1987 **D**

Polystira tellea (Dall, 1889)

Family Crassispiridae

Crassispira mesoleuca Rehder, 1943

Crassispira rhythmica Melvill, 1927

Hindsiclava alesidota (Dall, 1889)

Family Strictispiridae

Strictispira redferni Tippett, 2006

Strictispira solida (C.B. Adams, 1850)

Family Cochlespiridae

Cochlespira elegans (Dall, 1881)

Cochlespira radiata (Dall, 1889)
Leucosyrinx subgrundiferum (Dall, 1880)
Leucosyrinx tenoceras Dall, 1889
Leucosyrinx verrillii (Dall, 1881)
Pyrgospira ostrearum (Stearns, 1872)
Pyrgospira tampaensis (Bartsch and Rehder, 1939)

Family Clathurellidae

Bathytoma viabrunnea (Dall, 1889)
Drilliola loprestiana (Calcara, 1841) (*Microdrillia comatotropis* Dall is a synonym)
Glyphostoma dentifera Gabb, 1872
Glyphostoma gabbi Dall, 1889
Glyphostoma gratula (Dall, 1881)
Glyphostoma pilsbryi (Schwengel, 1940)
Lioglyphostoma adematum Woodring, 1928
Lioglyphostoma oenoa Bartsch, 1934
Miraclathurella herminea Bartsch, 1934
Mitrolumna biplicata Dall, 1889
Mitrolumna haycocki (Dall and Bartsch, 1911)
Nannodiella melanitica (Bush, 1885)
Nannodiella oxia (Bush, 1885)
Nannodiella pauca Fargo, 1953
Nannodiella vespuciana (d'Orbigny, 1847)
Tenaturris inepta (E.A. Smith, 1882)

Family Zonulispiridae

Compsodrillia acestra (Dall, 1889)
Compsodrillia disticha Bartsch, 1934
Compsodrillia eucosmia (Dall, 1889)
Compsodrillia haliostrephis (Dall, 1889)
Pilsbryspira albocincta (C.B. Adams, 1845)
Pilsbryspira jayana (C.B. Adams, 1850)
Pilsbryspira leucocyma (Dall, 1883)
Pilsbryspira monilis (Bartsch and Rehder, 1939)
Pilsbryspira nodata (C.B. Adams, 1850) (*albomaculata* is a synonym)
Zonulispira crocata (Reeve, 1845) (*Crassispira sanibelensis* is a synonym)

Family Drilliidae

Bellaspira pentagonalis (Dall, 1889)
Cerodrillia clappi Bartsch and Rehder, 1939 **E**
Cerodrillia hendersoni Bartsch, 1943
Cerodrillia perryae Bartsch and Rehder, 1939
Cerodrillia schroederi Bartsch and Rehder, 1939
Cerodrillia thea (Dall, 1884)
Cerodrillia williami Bartsch, 1943
Drillia acrybia (Dall, 1889)
Drillia pharcida (Dall, 1889)
Fenimorea fucata (Reeve, 1845)
Fenimorea halidorema Schwengel, 1940
Fenimorea kathyae Tippett, 1995 (north of the Dry Tortugas)
Fenimorea pagodula (Dall, 1889)
Fenimorea petiti Tippett, 1995 (north of the Dry Tortugas)

Fenimorea sunderlandi (Petuch, 1987) **D**
Leptodrillia splendida (Bartsch, 1934)
Neodrillia blacki Petuch, 2004
Neodrillia brunnescens (Rehder, 1943)
Neodrillia cydia Bartsch, 1943
Neodrillia moseri (Dall, 1889)
Neodrillia wolfei (Tippett, 1995)
Neodrillia woodringi (Bartsch, 1943)

Family Horaiclavidae
Inodrillia acova Bartsch, 1943
Inodrillia avira Bartsch, 1943
Inodrillia dido Bartsch, 1943
Inodrillia gibba Bartsch, 1943
Inodrillia hesperia Bartsch, 1943
Inodrillia hilda Bartsch, 1943
Inodrillia ino Bartsch, 1943
Inodrillia miamia Bartsch, 1943
Inodrillia vetula Bartsch, 1943

Family Raphitomidae
Anticlinura toreumata (Dall, 1889)
Brachycythara barbarae Lyons, 1972
Brachycythara biconica (C.B. Adams, 1850)
Cryoturris cerinella (Dall, 1889)
Cryoturris fargoi McGinty, 1955
Cryoturris filifera (Dall, 1881)
Cryoturris quadrilineata (C.B. Adams, 1850)
Daphnella corbicula Dall, 1889
Daphnella elata Dall, 1889
Daphnella eugrammata (Dall, 1902)
Daphnella lymnaeiformis (Kiener, 1840)
Daphnella margaretae Lyons, 1972
Daphnella retifera Dall, 1889
Daphnella stegeri McGinty, 1955
Eubela mcgintyi Schwengel, 1943
Glyphoturris quadrata (Reeve, 1845)
Gymnobella agassizii (Verrill and Smith, 1880)
Gymnobela blakeana (Dall, 1889)
Ithycthara auberiana (d'Orbigny, 1847)
Ithycythara lanceolata (C.B. Adams, 1850)
Ithycythara parkeri Abbott, 1958
Kurtziella accincta (Montagu, 1808)
Kurtziella atrostyla (Tryon, 1884)
Kurtziella limonitella (Dall, 1883)
Kurtziella perryae Bartsch and Rehder, 1939
Kurtziella serga (Dall, 1881)
Mangelia astricta Reeve, 1846
Mangelia bartletti (Dall, 1889)
Mangelia exsculpta (Watson, 1881)
Mangelia pelagia (Dall, 1881)

Mangelia pourtalesi (Dall, 1881)
Mangelia scipio (Dall, 1889)
Mangelia subsida (Dall, 1881)
Pyrgocythara balteata (Reeve, 1846)
Pyrgocythara candidissima (C.B. Adams, 1845)
Pyrgocythara densistriata (C.B. Adams, 1850)
Pyrgocythara filosa Rehder, 1939
Pyrgocythara hemphilli Bartsch and Rehder, 1939
Pyrgocythara plicosa (C.B. Adams, 1850)
Rimosodaphnella morra (Dall, 1881)
Rubellatoma diomedea Bartsch and Rehder, 1939
Rubellatoma rubella (Kurtz and Stimpson, 1851)
Stellatoma antonia (Dall, 1881)
Stellatoma monocingulata (Dall, 1889)
Stellatoma stellata (Stearns, 1872)

Order Heterobranchia
Superfamily Architectonicoidea
Family Architectonicidae
Architectonica nobilis Röding, 1798
Architectonica peracuta (Dall, 1889)
Architectonica sunderlandi Petuch, 1987
Heliacus bisulcatus (d'Orbigny, 1842)
Heliacus cylindrus (Gmelin, 1791)
Heliacus perrieri (Rochebrune, 1881)
Philippia krebsi (Mörch, 1875)
Solatisonax borealis (Verrill and Smith, 1880)

Family Mathildidae
Mathilda barbadensis Dall, 189
Mathilda hendersoni Dall, 1927
Mathilda yucatecana Dall, 1927

Superfamily Pyramidelloidea
Family Amathinidae
Amathina pacei (Petuch, 1987) **E**

Subclass Opisthobranchia
Order Cephalaspidea (groups with large shells only)
Superfamily Acteonoidea
Family Acteonidae
Acteon candens Rehder, 1939
Acteon cumingii A. Adams, 1854
Acteon danaida Dall, 1881
Acteon delicatus Dall, 1881
Acteon finlayi McGinty, 1955
Acteon incisus Dall, 1881
Acteon melampoides Dall, 1881
Acteon perforatus Dall, 1881
Acteon pusillus (Forbes, 1844)
Japonacteon punctostriata (C.B. Adams, 1840)
Ovulacteon meeki Dall, 1889

Family Bullinidae
  Bullina exquisita McGinty, 1955 **P**
Family Aplustridae
  Hydatina vesicaria (Lightfoot, 1786)
  Micromelo undatus (Bruguiere, 1792)
Superfamily Philinoidea
Family Cylichnidae
  Cylichnella bidentata (d'Orbigny, 1841)
  Scaphander punctostriatus Mighels, 1841
  Scaphander watsoni Dall, 1881
  Scaphander (Sabatia) bathymophila Dall, 1881
Superfamily Haminoeoidea
Family Haminoeidae
  Atys macandrewi E.A. Smith, 1872
  Atys riiseanus Mörch, 1875
  Atys sandersoni Dall, 1881
  Haminoea antillarum (d'Orbigny, 1841)
  Haminoea elegans (Gray, 1825)
  Haminoea succinea (Conrad, 1846)
  Haminoea taylorae Petuch, 1987 **E**
Family Acteocinidae
  Acteocina bullata (Kiener, 1834)
  Acteocina candei (d'Orbigny, 1842)
  Acteocina inconspicua Olsson and McGinty, 1958
  Acteocina recta (d'Orbigny, 1841)
  Utriculastra canaliculata (Say, 1822)
Superfamily Bulloidea
Family Bullidae
  Bulla eburnea Dall, 1881
  Bulla frankovichi Petuch and Sargent, 2011
  Bulla occidentalis A. Adams, 1850
  Bulla solida Gmelin, 1791
Subclass Pulmonata
Order Basommatophora
Superfamily Siphonarioidea
Family Siphonariidae
  Siphonaria (Patellopsis) alternata Say, 1826
  Siphonaria (Patellopsis) pectinata (Linnaeus, 1758)
  Williamia krebsi (Mörch, 1877)
Order Eupulmonata
Suborder Actophila
Superfamily Ellobioidea
Family Ellobiidae
  Apodopsis novimundi Pilsbry and McGinty, 1949
  Blauneria heteroclita (Montagu, 1808)
  Detracia bulloides (Montagu, 1808)
  Detracia floridana (Pfeiffer, 1856)
  Ellobium (Auriculoides) dominicense (Ferrusac, 1821)

Laemodonta cubensis (Pfeiffer, 1854)
Marinula succinea (Pfeiffer, 1854)
Melampus bidentatus (Say, 1822)
Melampus coffeus (Linnaeus, 1758)
Melampus monilis (Bruguiere, 1792)
Melampus morrisoni Martins, 1996
Microtralia occidentalis (Pfeiffer, 1854)
Ovatella myosotis (Draparnaud, 1801)
Pedipes mirabilis (Megerle von Mühlfeld, 1816)
Pedipes ovalis (C.B. Adams, 1849)
Tralia ovula (Bruguiere, 1792)

Order Stylommatophora
Suborder Sigmurethra
Superfamily Urocoptoidea
Family Cerionidae (salt-tolerant terrestrial species found in the supratidal zone)
Cerion incanum (Binney, 1851)
Cerion incanum fasciatum (Binney, 1859)
Cerion incanum saccharimeta Pilsbry and Vanatta, 1899
Cerion incanum vaccinum Pilsbry, 1902

Superfamily Orthalicoidea
Family Orthalicidae (arboreal gastropods with color forms and subspecies endemic to the Keys)
Liguus fasciatus castaneozonatus Pilsbry, 1912 (with 2 named color forms in the Keys)
Liguus fasciatus elliottensis Pilsbry, 1912 (with 1 named color form in the Keys)
Liguus fasciatus graphicus Pilsbry, 1912 (with 8 named color forms in the Keys)
Liguus fasciatus lossmanicus Pilsbry, 1912 (with 2 named color forms in the Keys)
Liguus fasciatus matecumbensis Pilsbry, 1912 (with 1 named color form in the Keys)
Liguus fasciatus solidus (Say, 1825) (with 4 named color forms in the Keys)
Liguus fasciatus testudineus Pilsbry, 1912 (with 2 named color forms in the Keys)
Orthalicus floridensis (Pilsbry, 1899)
Orthalicus reses (Say, 1825)
Orthalicus reses nesodryas Pilsbry, 1946

Family Bulimulidae (arboreal gastropods)
Drymaeus hemphilli (B. H. Wright, 1889)
Drymaeus multilineatus (Say, 1825)
Drymaeus multilineatus latizonatus Pilsbry, 1936

Class Bivalvia
Subclass Protobranchia
Order Nuculoida
Superfamily Nuculoidea
Family Nuculidae
Eunucula aegeensis (Forbes, 1844)
Nucula calcicola D. Moore, 1977
Nucula crenulata A. Adams, 1856
Nucula delphinodonta Mighels and C.B. Adams, 1842
Nucula proxima Say, 1822

Order Solemyoida
Superfamily Solemyoidea

Family Solemyidae
  Solemya occidentalis Deshayes, 1857
  Solemya velum Say, 1822
Order Nuculanoida
Superfamily Nuculanoidea
Family Nuculanidae
  Ledella sublaevis Verrill and Bush, 1898
  Nuculana acuta (Conrad, 1832)
  Nuculana concentrica (Say, 1824)
  Nuculana jamaicensis (d'Orbigny, 1853)
  Nuculana semen (E.A. Smith, 1885)
  Nuculana verrilliana (Dall, 1886)
  Nuculana vitrea (d'Orbigny, 1853)
  Propeleda carpenteri (Dall, 1881)
Family Yoldiidae
  Yoldia liorhina Dall, 1881
Subclass Pteriomorpha
Order Arcoida
Superfamily Arcoidea
Family Arcidae
  Acar domingensis (Lamarck, 1819)
  Anadara baughmani (Hertlein, 1951)
  Anadara floridana (Conrad, 1869)
  Anadara transversa (Say, 1822)
  Arca imbricata Bruguiere, 1792
  Arca rachelcarsonae Petuch and Myers, new species **E**
  Arca zebra (Swainson, 1833)
  Barbatia cancellaria (Lamarck, 1811)
  Bathyarca glomerula (Dall, 1881)
  Bentharca sagrinata (Dall, 1886)
  Caloosarca notabilis (Röding, 1798)
  Cucullearca candida (Helbling, 1779)
  Fugleria tenera (C.B. Adams, 1845)
  Lunarca ovalis (Bruguiere, 1789)
  Scapharca brasiliana (Lamarck, 1819)
  Scapharca chemnitzii (Philippi, 1851)
Family Noetiidae
  Arcopsis adamsi (Dall, 1886)
  Noetia ponderosa (Say, 1822)
Family Glycymeridae
  Glycymeris americana (DeFrance, 1826)
  Glycymeris decussata (Linnaeus, 1758)
  Glycymeris spectralis Nicol, 1952
  Glycymeris undata (Linnaeus, 1758)
  Tucetona pectinata (Gmelin, 1791)
  Tucetona subtilis Nicol, 1956
Superfamily Limopsoidea
Family Limopsidae
  Limopsis aurita (Brocchi, 1814)

Limopsis cristata Jeffreys, 1876
Limopsis minuta (Philippi, 1836)
Limopsis sulcata Verrill and Bush, 1898

Order Mytiloida
Superfamily Mytiloidea
Family Mytilidae

Amygdalum politum (Verrill and Smith, 1880)
Amygdalum sagittatum (Rehder, 1935)
Arcuatula papyria (Conrad, 1846)
Brachidontes domingensis (Lamarck, 1819)
Brachidontes exustus (Linnaeus, 1758)
Brachidontes modiolus (Linnaeus, 1758)
Crenella decussata (Montagu, 1808)
Dacrydium hendersoni Salas and Gofas, 1997
Geukensia granosissima (Sowerby III, 1914)
Gregariella coralliophaga (Gmelin, 1791)
Ischadium recurvum (Rafinesque, 1820)
Lioberus castaneus (Say, 1822)
Lithophaga antillarum (d'Orbigny, 1853)
Lithophaga aristata (Dillwyn, 1817)
Lithophaga bisulcata (d'Orbigny, 1853)
Lithophaga nigra (d'Orbigny, 1853)
Modiolus americanus (Leach, 1815)
Modiolus squamosus Beauperthuy, 1967
Musculus luteralis (Say, 1822)

Order Pterioida
Superfamily Pterioidea
Family Pteriidae

Pinctada imbricata Röding, 1798
Pinctada longisquamosa (Dunker, 1852)
Pteria colymbus (Röding, 1798)
Pteria vitrea (Reeve, 1857)

Family Isognomonidae

Isognomon alatus (Gmelin, 1791)
Isognomon bicolor (C.B. Adams, 1845)
Isognomon radiatus (Anton, 1838)

Family Malleidae

Malleus candeanus (d'Orbigny, 1853)

Superfamily Ostreoidea
Family Ostreidae

Crassostrea rhizophorae (Guilding, 1828)
Crassostrea virginica (Gmelin, 1791)
Dendrostrea frons (Linnaeus, 1758)
Ostreola equestris (Say, 1834)
Teskeyostrea weberi (Olsson, 1951)

Family Gryphaeidae

Hyotissa mcgintyi (Harry, 1985)
Neopycnodonte cochlear (Poli, 1795)

Superfamily Pinnoidea

Family Pinnidae
> *Atrina rigida* (Lightfoot, 1786)
> *Atrina seminuda* (Lamarck, 1786)
> *Atrina serrata* (Sowerby I, 1825)
> *Pinna carnea* Gmelin, 1791

Order Limoida
Superfamily Limoidea
Family Limidae
> *Ctenoides mitis* (Lamarck, 1807)
> *Ctenoides planulata* (Dall, 1886)
> *Ctenoides samanensis* Stuardo, 1982 (*miamiensis* is a synonym)
> *Ctenoides sanctipauli* Stuardo, 1982
> *Ctenoides scabra* (Born, 1778)
> *Divarilima albicoma* (Dall, 1886)
> *Lima caribaea* d'Orbigny, 1853
> *Limaria pellucida* (C.B. Adams, 1846)
> *Limatula confusa* (E.A. Smith, 1885)
> *Limatula setifera* Dall, 1886
> *Limatula subovata* (Jeffreys, 1876)
> *Limea bronniana* Dall, 1886

Order Pectinoida
Superfamily Pectinoidea
Family Pectinidae
> *Aequipecten glyptus* (Verrill, 1882)
> *Aequipecten lineolaris* (Lamarck, 1819)
> *Antillipecten antillarum* (Recluz, 1853)
> *Argopecten gibbus* (Linnaeus, 1758)
> *Argopecten irradians taylorae* Petuch, 1987 (Florida Keys; western Florida to Mississippi)
> *Argopecten nucleus* (Born, 1778)
> *Caribachlamys imbricatus* (Gmelin, 1791) (incorrectly referred to as *pellucens*)
> *Caribachlamys mildredae* (Bayer, 1941)
> *Caribachlamys ornatus* (Lamarck, 1819)
> *Caribachlamys sentis* (Reeve, 1853)
> *Cryptopecten phrygium* (Dall, 1886)
> *Euvola chazaliei* (Dautzenberg, 1900)
> *Euvola laurenti* (Gmelin, 1791)
> *Euvola marshallae* Petuch and Myers, new species (incorrectly referred to as *papyracea*)
> *Euvola raveneli* (Dall, 1898)
> *Euvola ziczac* (Linnaeus, 1758)
> *Laevichlamys multisquamata* (Dunker, 1864)
> *Lindapecten exasperatus* (Sowerby II, 1842) (*acanthodes* is a synonym)
> *Lindapecten lindae* Petuch, 1995
> *Lindapecten muscosus* (Wood, 1828)
> *Nodipecten fragosus* (Conrad, 1849)
> *Spathochlamys benedicti* (Verrill and Bush, 1897)

Family Propeamussiidae
> *Cyclopecten thalassinus* (Dall, 1886)
> *Hyalopecten strigillatus* (Dall, 1889)
> *Parvamussium pourtalesianum* (Dall, 1886)

    Parvamussium sayanum (Dall, 1886)
    Propeamussium cancellatum (E.A. Smith, 1885)
    Similipecten nanus (Verrill and Bush, 1897)
Family Spondylidae
    Spondylus americanus Hermann, 1781
    Spondylus tenuis Schriebers, 1793 (*ictericus* is a synonym)
Superfamily Plicatuloidea
Family Plicatulidae
    Plicatula gibbosa Lamarck, 1801
Superfamily Anomioidea
Family Anomiidae
    Anomia simplex d'Orbigny, 1853
    Pododesmus rudis (Broderip, 1834)
Subclass Heterodonta
Order Carditoida
Superfamily Crassitelloidea
Family Crassitellidae
    Crassinella dupliniana (Dall, 1903)
    Crassinella lunulata (Conrad, 1834)
    Crassinella martinicensis (d'Orbigny, 1853)
    Eucrassatella speciosa (A. Adams, 1852)
Family Astartidae
    Astarte nana E.A. Smith, 1881
    Astarte smithii Dall, 1866
    Astarte subuequilatera Sowerby II, 1854
Family Carditidae
    Carditamera floridana Conrad, 1838
    Glans dominguensis (d'Orbigny, 1853)
    Pleuromeris tridentata (Say, 1826)
    Pteromeris perplana (Conrad, 1841)
Order Anomalodesmata
Superfamily Pandoroidea
Family Pandoridae
    Pandora bushiana Dall, 1886
    Pandora glacialis Leach, 1819
    Pandora inflata Boss and Merrill, 1965
Family Lyonsiidae
    Entodesma beana (d'Orbigny, 1853)
    Lyonsia floridana Conrad, 1849
Family Periplomatidae
    Periploma margaritaceum (Lamarck, 1801)
    Periploma tenerum (Fischer, 1882)
Family Thraciidae
    Asthenothaerus hemphilli Dall, 1886
    Bushia elegans (Dall, 1886)
    Thracia morrisoni Petit, 1964
    Thracia phaseolina Lamarck, 1911
    Thracia stimpsoni Dall, 1886
Order Septibranchia

Superfamily Poromyoidea
Family Poromyidae
 Poromya albida Dall, 1886
 Poromya granulata (Nyst and Westendorp, 1839)
 Poromya rostrata Rehder, 1943
Family Verticordia
 Euciroa elegantissima (Dall, 1881)
 Haliris fischeriana (Dall, 1881)
 Spinosipella acuticostata (Philippi, 1844)
 Trigonulina ornata d'Orbigny, 1853
Family Cuspidariidae
 Cardiomya alternata (d'Orbigny, 1853)
 Cardiomya costellata (Deshayes, 1833)
 Cardiomya gemma Verrill and Bush, 1898
 Cardiomya perrostrata (Dall, 1881)
 Cardiomya ornatissima (d'Orbigny, 1853)
 Cardiomya striata (Jeffreys, 1876)
 Cuspidaria obesa (Loven, 1846)
 Cuspidaria rostrata (Spengler, 1793)
 Myonera gigantea (Verrill, 1884)
 Myonera lamellifera (Dall, 1881)
 Myonera paucistriata Dall, 1886
 Plectodon granulatus (Dall, 1881)
Order Veneroida
Superfamily Lucinoidea
Family Lucinidae
 Anodontia alba Link, 1807
 Anodontia schrammi (Crosse, 1876)
 Callucina keenae Chavan, 1971
 Cavilinga blanda (Dall, 1901)
 Codakia orbicularis (Linnaeus, 1758)
 Ctena orbiculata (Montagu, 1808)
 Ctena pectinella (C.B. Adams, 1852)
 Divalinga quadricostata (d'Orbigny, 1845)
 Divaricella dentata (Wood, 1815)
 Lucina pensylvanica (Linnaeus, 1758)
 Lucinisca muricata (Spengler, 1798)
 Lucinisca nassula (Conrad, 1846)
 Myrtea sagrinata (Dall, 1886)
 Myrteopsis lens (Verrill and Smith, 1880)
 Parvilucina costata (d'Orbigny, 1845)
 Parvilucina crenella (Dall, 1901)
 Phacoides pectinata (Gmelin, 1791)
 Pleurolucina leucocyma (Dall, 1886)
 Pleurolucina sombrerensis (Dall, 1886)
 Radiolucina amianta (Dall, 1901)
 Stewartia floridana (Conrad, 1833)
Family Ungulinidae
 Diplodonta notata Dall and Simpson, 1901

    Diplodonta nucleiformis (Wagner, 1840)
    Diplodonta punctata (Say, 1822)
    Phlyctiderma semiaspera (Philippi, 1836)
    Phlyctiderma soror (C.B. Adams, 1852)
Family Thyasiridae
    Thyasira grandis (Verrill and Smith, 1885)
    Thyasira trisinuata (d'Orbigny, 1853)
Superfamily Chamoidea
Family Chamidae
    Arcinella cornuta Conrad, 1866
    Chama congregata Conrad, 1833
    Chama florida Lamarck, 1819
    Chama inezae (Bayer, 1943)
    Chama lactuca Dall, 1886
    Chama macerophylla Gmelin, 1791
    Chama radians Lamarck, 1819
    Chama sarda Reeve, 1847
    Chama sinuosa Broderip, 1835
Superfamily Galeommatoidea
Family Lasaeidae
    Erycina periscopiana Dall, 1899
    Kellia suborbicularis (Montagu, 1803)
    Lasaea adansoni (Gmelin, 1791)
    Mysella planulata (Stimpson, 1851)
    Orobitella floridana (Dall, 1899)
Superfamily Hiatelloidea
Family Hiatellidae
    Hiatella arctica (Linnaeus, 1758)
    Hiatella azaria (Dall, 1881)
Superfamily Gastrochaenoidea
Family Gastrochaenidae
    Gastrochaena ovata (Sowerby I, 1834)
    Lamychaena hians (Gmelin, 1791)
    Spengleria rostrata (Spengler, 1783)
Superfamily Arcticoidea
Family Trapezidae
    Coralliophaga coralliophaga (Gmelin, 1791)
Superfamily Cyamioidea
Family Sportellidae
    Basterotia elliptica (Recluz, 1850)
    Basterotia quadrata (Hinds, 1843)
    Ensitellops protexta (Conrad, 1841)
Superfamily Sphaerioidea
Family Corbiculidae
    Polymesoda maritima (d'Orbigny, 1842) (*P. floridana* is a synonym)
Superfamily Cardioidea
Family Cardiidae
    Acrosterigma magnum (Linnaeus, 1758)
    Americardia guppyi Theile, 1910

Americardia lightbourni Lee and Huber, 2012
Americardia media (Linnaeus, 1758)
Dallocardia muricata (Linnaeus, 1758)
Dinocardium vanhyningi Clench and Smith, 1944 **T**
Laevicardium mortoni (Conrad, 1830)
Laevicardium pictum (Ravenel, 1861)
Laevicardium serratum (Linnaeus, 1758)
Microcardium peramabile (Dall, 1881)
Microcardium tinctum (Dall, 1881)
Papyridea lata (Born, 1778)
Papyridea semisulcata (Gray, 1825)
Papyridea soleniformis (Bruguiere, 1789)
Trachycardium egmontianum (Shuttleworth, 1856)
Trigonocardia antillarum (d'Orbigny, 1842)

Superfamily Veneroidea
Family Veneridae
Anomalocardia cuneimeris (Conrad, 1846)
Callista eucymata (Dall, 1890)
Callista maculata (Linnaeus, 1758)
Chione cancellata (Linnaeus, 1758)
Chione elevata Say, 1822
Chione mazyckii Dall, 1902
Chionopsis intapurpurea (Conrad, 1849)
Choristodon robustum (Sowerby I, 1834)
Circomphalis strigillinus (Dall, 1902)
Cooperella atlantica Rehder, 1943
Cyclinella tenuis (Recluz, 1852)
Dosinia discus (Reeve, 1850)
Dosinia elegans (Conrad, 1846)
Gemma gemma (Totten, 1834)
Globivenus rigida (Dillwyn, 1817)
Globivenus rugatina (Heilprin, 1886)
Gouldia cerina (C.B. Adams, 1845)
Lirophora clenchi (Pulley, 1952)
Lirophora paphia (Linnaeus, 1758)
Lirophora varicosa (Sowerby II, 1853) (= *latilirata* of authors)
Macrocallista nimbosa (Lightfoot, 1786)
Mercenaria campechiensis (Gmelin, 1791)
Mercenaria hartae Petuch, 2013 **P**
Mercenaria mercenaria notata (Say, 1822) **P**
Parastarte triquetra (Conrad, 1846)
Periglypta listeri (Gray, 1838)
Petricola lapicida (Gmelin, 1791)
Petricolaria pholadiformis (Lamarck, 1818)
Pitar albidus (Gmelin, 1791)
Pitar circinatus (Born, 1778)
Pitar dione (Linnaeus, 1758)
Pitar fulminatus (Menke, 1828)
Pitar simpsoni (Dall, 1895)

Pitarenus cordatus (Schwengel, 1951)
Puberella pubera (Bory de Saint-Vincent, 1827)
Timoclea grus (Holmes, 1858)
Timoclea pygmaea (Lamarck, 1818)
Tivela abaconis Dall, 1902
Tivela floridana Rehder, 1939 **P**
Tivela mactroides (Born, 1778)
Transennella conradiana (Dall, 1884)
Transennella cubaniana (d'Orbigny, 1853)
Transennella culebrana (Dall and Simpson, 1901)
Transennella stimpsoni Dall, 1902

Superfamily Tellinoidea
Family Tellinidae

Acorylus gouldi (Hanley, 1846)
Angulus agilis (Stimpson, 1857)
Angulus merus (Say, 1834)
Angulus paramerus (Boss, 1964)
Angulus probinus (Boss, 1964)
Angulus sybariticus (Dall, 1881)
Angulus tampaensis (Conrad, 1866)
Angulus texanus (Dall, 1900)
Angulus versicolor (DeKay, 1843)
Arcopagia fausta (Pulteney, 1799)
Cymatoica hendersoni Rehder, 1939
Elliptotellina americana (Dall, 1900)
Eurytellina angulosa (Gmelin, 1791)
Eurytellina alternata (Say, 1822)
Eurytellina lineata (Turton, 1819)
Eurytellina nitens (C.B. Adams, 1845)
Eurytellina punicea (Born, 1778)
Laciolina laevigata (Linnaeus, 1758)
Laciolina magna (Spengler, 1798)
Leporimetus intastriata (Say, 1826)
Macoma brevifrons (Say, 1834)
Macoma cerina (C.B. Adams, 1845)
Macoma constricta (Bruguiere, 1792)
Macoma extenuata Dall, 1900
Macoma limula Dall, 1895
Macoma pseudomera Dall and Simpson, 1901
Macoma tageliformis Dall, 1900
Macoma tenta (Say, 1834)
Merisca aequistriata (Say, 1824)
Merisca cristallina (Spengler, 1798)
Merisca martinicensis (d'Orbigny, 1853)
Phyllodina squamifera (Deshayes, 1855)
Scissula candeana (d'Orbigny, 1853)
Scissula consobrina (d'Orbigny, 1853)
Scissula iris (Say, 1822)
Scissula similis (Sowerby I, 1806)

Strigilla carnaria (Linnaeus, 1758)
Strigilla mirabilis (Philippi, 1841)
Strigilla pisiformis (Linnaeus, 1758)
Tellidora cristata (Recluz, 1842)
Tellina radiata (Linnaeus, 1758)
Tellinella listeri (Röding, 1798)

Family Donacidae
Donax variabilis Say, 1822
Iphigenia brasiliana (Lamarck, 1818)

Family Psammobiidae
Asaphis deflorata (Linnaeus, 1758)
Gari circe (Mörch, 1876)
Heterodonax bimaculatus (Linnaeus, 1758)
Sanguinolaria sanguinolenta (Gmelin, 1791)

Family Semelidae
Abra aequalis (Say, 1822)
Abra americana Verrill and Bush, 1898
Abra lioica (Dall, 1881)
Cumingia coarctata Sowerby I, 1833
Cumingia vanhyningi Rehder, 1939
Ervilia concentrica (Holmes, 1858)
Ervilia nitens (Montagu, 1808)
Ervilia subcancellata E.A. Smith, 1885
Semele bellastriata (Conrad, 1837)
Semele donovani McGinty, 1955 **P**
Semele proficua (Pulteney, 1799)
Semele purpurascens (Gmelin, 1791)
Semelina nuculoides (Conrad, 1841)

Family Solecurtidae
Solecurtus cumingianus (Dunker, 1861)
Tagelus divisus (Spengler, 1794)
Tagelus plebeius (Lightfoot, 1786)

Superfamily Solenoidea
Family Pharidae
Ensis minor Dall, 1900

Superfamily Mactroidea
Family Mactridae
Anatina anatina (Spengler, 1802)
Mactrotoma fragilis (Gmelin, 1791)
Mulinia lateralis (Say, 1822)
Raeta plicatella (Lamarck, 1818)
Spisula raveneli (Conrad, 1832)

Superfamily Dreissenoidea
Family Dreissenidae
Mytilopsis leucophaeata (Conrad, 1831)
Mytilopsis sallei (Recluz, 1849)

Superfamily Myoidea
Family Myidae
Sphenia fragilis (H. and A. Adams, 1854)

Family Corbulidae

Caryocorbula caribaea (d'Orbigny, 1853)
Caryocorbula chittyana (C.B. Adams, 1852)
Caryocorbula contracta (Say, 1822)
Caryocorbula cymella (Dall, 1881)
Caryocorbula dietziana (C.B. Adams, 1852)
Juliacorbula aequivalvis (Philippi, 1836)
Variacorbula limatula (Conrad, 1846)
Variacorbula philippii (E.A. Smith, 1845)

Superfamily Pholadoidea

Family Pholadidiae

Barnea truncata (Say, 1822)
Cyrtopleura costata (Linnaeus, 1758)
Martesia cuneiformis (Say, 1822)
Martesia striata (Linnaeus, 1758)
Pholas campechiensis Gmelin, 1791

View of a small Red Mangrove and Oyster (*Crassostrea virginica*) island in Chokoloskee Bay, Collier County, Florida. This island formation, composed of a combination of oyster bioherms and mangrove forests, is typical of the Inner Ten Thousand Islands of the Reticulated Coastal Swamps.

Ecologic, oceanographic, and geographic terms

The following is a list of ecological, geographical, geomorphological, and oceanographic terms that appear throughout the book. Only those that refer directly to the Florida Keys and adjacent areas are included.

A

Aggregated Patch Reef Macrohabitat
Alligator Reef
Anastasia Formation
Arca rachelcarsonae Assemblage
Atlantic Coastal Ridge

B

Back Bays
Bahia Honda Key
Barnes Sound
Batillaria minima Assemblage
Bayericerithium litteratum Assemblage
Bermont Formation
Big Cypress Swamp
Big Pine Key
Big Pine-Little Torch Channel
Biscayne Bay
Black Mangrove Forests
Black Water Sound
Boca Grande Key
Broward County
Bulla occidentalis Assemblage

C

Caloosahatchee Formation
Cape Romano
Cape Sable
Card Sound
Caribbean Current
Caribbean Molluscan Province
Carolinian Molluscan Province
Carysfort Reef
Cenchritis muricatus Assemblage

Cerithideopsis costatus Assemblage
Cerithium lutosum Assemblage
Cerodrillia clappi Assemblage
Chicoreus rachelcarsonae Assemblage
Chokoloskee Bay, Ten Thousand Islands
Chokoloskee Pass, Ten Thousand Islands
Coastal Marine Ecological Classification Standard (CMECS)
Coffee Mill Hammock Member
Collier County
Comer Key, Ten Thousand Islands
Conch Key
Coralline Algal Beds
Crane Point Hammock
Crassostrea rhizophorae Assemblage
Cymatium rehderi Assemblage
Cyphoma rhomba Assemblage

D

Dade County
Deep Coralline Algal Beds
Deep Coral Reef Formation
Deep Reef Talus Macrohabitat
Deep Softbottom Macrohabitat
Deep Softbottom Terraces
Demes
Demijohn Key, Ten Thousand Islands
Distal Atolls
Dry Tortugas
Dwarf Mangrove Meadows

E

Elliott Key
Estuarine Regime
Everglades National Park
Everglades Pseudoatoll

Dorsal view of *Dauciconus glicksteini* (Petuch, 1987), one of the rarest cone shells from the deep coral reefs off southeastern Florida and the Florida Keys. This specimen was collected off Boynton Beach, Palm Beach County, Florida. (Photography by Jennie Petuch)

Illustrated specimens

The species and subspecies illustrated in Chapters 2 through 10 are listed here by figure number. Species that are listed, but not illustrated, in the individual biodiversity sections are not included.

A

Aliger gallus	Figure 8.3G, H
Amathina pacei	Figure 7.9C–E
Americardia lightbourni	Figure 9.3B
Americoliva bollingi	Figure 6.8A–F
Americoliva edwardsae	Figure 1.5I
Americoliva matchetti	Figure 9.6D–F
Americoliva recourti	Figure 9.6A–C
Americoliva sunderlandi	Figure 9.3J, K
Amerilittorina angustior	Figure 3.6G, H
Amerilittorina jamaicensis	Figure 3.6I, J
Amerilittorina placida	Figure 3.6L
Amerilittorina ziczac	Figure 3.6K
Angiola lineata	Figure 3.7B
Angulus sybariticus	Figure 6.13F
Anodontia alba	Figure 4.16C
Anomalocardia cuneimeris	Figure 6.3G
Anomia simplex	Figure 2.7E
Antillipecten antillarum	Figure 4.12I, J
Arca imbricata	Figure 5.8H
Arca rachelcursonae	Figure 8.4A–C
Arca zebra	Figure 5.8L
Architectonica peracuta	Figure 9.4D
Arcopagia fausta	Figure 4.13A
Arcopsis adamsi	Figure 3.10F
Arcuatula papyria	Figure 4.5A
Arene cruentata	Figure 7.3A
Arene vanhyningi	Figure 7.3B
Argopecten gibbus	Figure 6.7J
Argopecten irradians taylorae	Figure 4.12A–D
Argopecten nucleus	Figure 4.12E, F
Asaphis deflorata	Figure 6.2K
Astralium phoebia	Figure 4.8B
Atlanticonus granulatus	Figure 8.2J
Atrina rigida	Figure 4.16J
Attenuiconus attenuatus	Figure 6.9B
Attiliosa philippiana	Figure 8.4L
Aurinia dubia	Figure 9.5A, B

B

Babelomurex scalariformis	Figure 7.4E
Bailya intricata	Figure 7.4B
Bailya parva	Figure 7.4H
Barbatia cancellaria	Figure 3.10E
Burnea truncata	Figure 4.5K
Bartschia significans	Figure 9.4F
Batillaria minima	Figure 6.2C, D
Bayericerithium litteratum	Figure 5.8C, D
Bayericerithium semiferrugineum	Figure 7.4J
Bostrycapulus aculeatus	Figure 5.8I
Brachidontes exustus	Figure 2.7D
Brachidontes modiolus	Figure 5.2L
Bulla frankovichi	Figure 5.4D, E, L
Bulla occidentalis	Figure 4.2J, K
Bursa cubaniana	Figure 7.6J
Bursa (Colubrellina) corrugata	Figure 7.6F, G
Bursa (Colubrellina) tenuisculpta	Figure 8.4E
Bursa (Lampasopsis) grayana	Figure 7.6I

C

Calliostoma adelae	Figure 4.8D
Calliostoma fascinans	Figure 8.2J
Calliostoma javanicum	Figure 7.3G
Calliostoma jujubinum	Figure 7.3D
Calliostoma (Kombologion) benedicti	Figure 8.4D
Calliostoma pulchrum	Figure 7.3C
Calliostoma scalenum	Figure 8.3D
Calliostoma tampaense	Figure 7.6L
Callista maculata	Figure 6.7H
Caloosarca notabilis	Figure 6.3I

Index

View of a Florida Keys sunset, looking across the mangrove forests on No Name Key, Lower Florida Keys. (Photograph by Rob Myers)

T - #0504 - 071024 - C320 - 254/178/14 - PB - 9780367658915 - Gloss Lamination